BORNOLOGIES AND FUNCTIONAL ANALYSIS

W9-AEB-811

NORTH-HOLLAND
MATHEMATICS STUDIES **26**

Notas de Matemática (62)
Editor: Leopoldo Nachbin
Universidade Federal do Rio de Janeiro
and University of Rochester

BORNOLOGIES AND FUNCTIONAL ANALYSIS

Introductory course on the theory of duality topology-bornology and its use in functional analysis

Henri Hogbe-Nlend
Professor of Mathematics
University of Bordeaux, France
and
Directeur de recherches
Laboratoire Associé 226 du C.N.R.S.

Translated from the French
by
V.B. Moscatelli, University of Sussex

1977

NORTH-HOLLAND PUBLISHING COMPANY – AMSTERDAM • NEW YORK • OXFORD

© North-Holland Publishing Company – 1977

All rights reserved. No part of this publication may be reproduced, stored in a retrieval system, or transmitted, in any form or by any means, electronic, mechanical, photocopying, recording or otherwise, without the prior permission of the copyright owner.

North-Holland ISBN: 0 7204 0712 5

PUBLISHERS:
NORTH-HOLLAND PUBLISHING COMPANY
AMSTERDAM • NEW YORK • OXFORD

SOLE DISTRIBUTORS FOR THE U.S.A. AND CANADA:
ELSEVIER NORTH-HOLLAND, INC.
52 VANDERBILT AVENUE, NEW YORK, N.Y. 10017

Library of Congress Cataloging in Publication Data

Hogbe-Nlend, H
Bornologies and functional analysis.

(Notas de matemática;
Bibliography: p. 62
Includes index.
1. Functional analysis. 2. Bornological spaces. 3. Duality theory (Mathematics) 4. Differential equations, Partial. I. Title. II. Series.
QA1.N86 [QA320] 510'.8s [515'.7] 77-815
ISBN 0-7204-0712-5 (Elsevier)

PRINTED IN THE NETHERLANDS

Sep

1593449

INTRODUCTION

Modern Functional Analysis is the study of infinite-dimensional vector spaces and operators acting between these spaces, based upon the notion of convergence. The main ideas used are those of *locally convex topology* and of *convex bornology*. The present course gives, for the first time, an introductory exposition of the theory of Bornology and its use in Functional Analysis.

After a systematic account of the fundamental bornological notions, we study the deep duality relationships, internal and external, between topology and bornology, which enable us to present the fundamental classes of spaces in a new light: bornological, completely bornological or ultra-bornological, barrelled, reflexive, completely reflexive, hypo-Montel, Montel, Schwartz, co-Schwartz and Silva spaces. These spaces form the general and precise framework in which the fundamental theorems and techniques of Functional Analysis hold, and these theorems and techniques are established in this course in all the generality required by the applications. The last chapter, devoted to Partial Differential Equations, gives a concrete illustration of the general results obtained.

The present text is intended for undergraduate students (from the second year), étudiants du troisième cycle, and beginning research workers in the field of Functional Analysis; it originated in courses given by the author at the University of Bordeaux since 1968 and at the University of São-Paulo during 1972-1973.

It is a great pleasure for me to extend my sincere gratitude to Dr. V.B. Moscatelli of the University of Sussex for translating my French manuscript into English.

In the near future this book will be followed by another entitled *Nuclear and Co-Nuclear Spaces*.

H. HOGBE-NLEND

Bordeaux, January 1976

THE VARIOUS BRANCHES OF FUNCTIONAL ANALYSIS AND THEIR MUTUAL RELATIONSHIPS

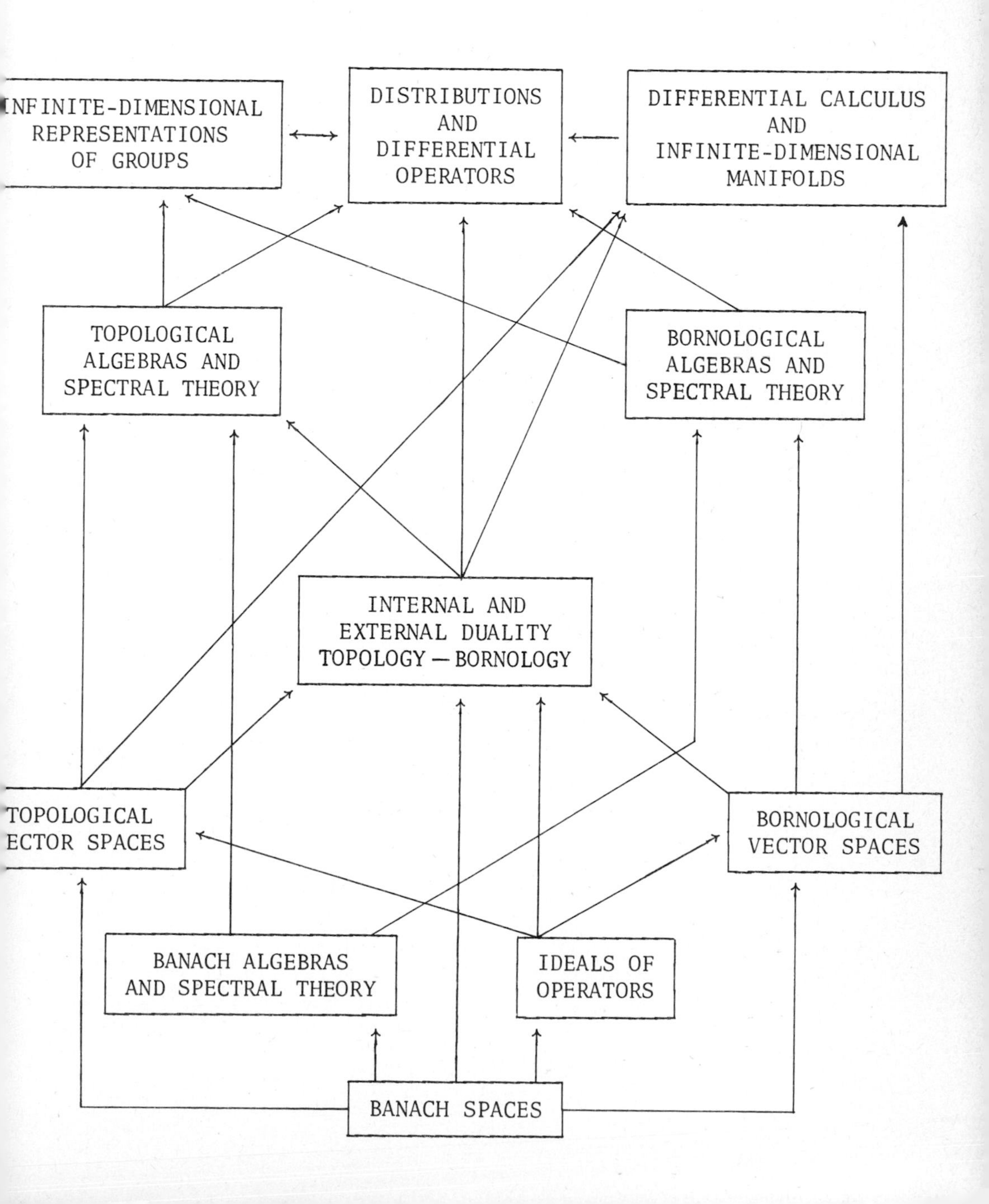

CONTENTS

DISTRIBUTIONS AND DIFFERENTIAL OPERATORS

EXERCISES

CHAPTER 0

PRELIMINARY NOTIONS
OF ALGEBRA AND TOPOLOGY

The essential character of the theory developed in this book is the *simultaneous* consideration of three structures on the same set: an algebraic structure (which will always be that of a vector space), a topological structure and a 'bornological structure'. The first two are classical and well known, and we shall only need elementary results from their theories, which we collect in this Chapter.

0·A VECTOR SPACES

0·A.0 PRELIMINARIES

For elementary set theory we follow the notation of the treatise by Dieudonné [*1*] unless the contrary is expressly stated.

We assume the reader to be familiar with the most elementary notions of linear algebra (*cf.*, for example, Dieudonné [*1*], Annexe). All vector spaces considered in this book are over the same field $\mathbb{K}$ which will always be the real field $\mathbb{R}$ or the complex field $\mathbb{C}$. We shall then speak sometimes of vector spaces without mentioning the field explicitly.

0·A.1 INDUCTIVE LIMITS OF VECTOR SPACES

In this paragraph I stands for a non-empty, ordered set of indices which is directed, i.e. for every pair $(i,j) \in I \times I$ there exists $k \in I$ such that $k \geq i$ and $k \geq j$.

0·A.1·1 Inductive Systems of Vector Spaces

Let $(E_i)_{i \in I}$ be a family of vector spaces over $\mathbb{K}$. Suppose that for every pair $(i,j) \in I \times I$ such that $i \leqslant j$, there exists a linear map $u_{ji}:E_i \to E_j$ such that the system of maps (u_{ji}) satisfies the following conditions:

(i) For every $i \in I$, $u_{ii}:E_i \to E_i$ is the identity map of E_i;

(ii) For every i,j,k elements of I such that $i \leqslant j \leqslant k$, we have $u_{ki} = u_{kj} \circ u_{ji}$.

The system (E_i, u_{ji}) is called an *INDUCTIVE SYSTEM OF VECTOR SPACES*.

0·A.1˙2 Existence and Uniqueness of the Inductive Limit of Vector Spaces

Let (E_i, u_{ji}) be an inductive system of vector spaces over $\mathbb{K}$. There exist a vector space E over $\mathbb{K}$ and, for each $i \in I$, a linear map $u_i:E_i \to E$, such that:

(IL.1): $u_i = u_j \circ u_{ji}$ *whenever* $i \leqslant j$;

(IL.2): *For every vector space* F *and family of linear maps* $v_i:E_i \to F$ *such that* $v_i = v_j \circ u_{ji}$ *for* $i \leqslant j$, *there exists a* unique *linear map* $v:E \to F$ *satisfying* $v_i = v \circ u_i$.

The vector space E is unique up to isomorphism and is called the *INDUCTIVE LIMIT* of the inductive system (E_i, u_{ji}). For every $i \in I$, the map $u_i:E_i \to E$ is called the *CANONICAL MAP* of E_i into E.

In practice knowledge of the proofs of the above statements is not useful, whilst the following properties are.

0·A.1˙3 Properties of Inductive Limits

(a): $E = \bigcup_{i \in I} u_i(E_i)$ *in the the set-theoretical sense.*

(b): *If for every* $i \leqslant j$ *the map* u_{ji} *is injective, then it is* u_i *for each* $i \in I$.

(c): *From the algebraic point of view the operations on* E *are defined as follows. ADDITION: if* $x \in E$ *and* $y \in E$, *there exist* $i \in I$, $x_i \in E_i$ *and* $y_i \in E_i$ *such that* $x = u_i(x_i)$ *and* $y = u_i(y_i)$, *and for every such system* (i, u_i, x_i, y_i) *we have:*

$$x + y = u_i(x_i + y_i).$$

SCALAR MULTIPLICATION is defined analogously: if $\lambda \in \mathbb{K}$ *and* $x \in E$, *there exist* $i \in I$ *and* $x_i \in E_i$ *such that* $x = u_i(x_i)$, *and for every such system we have:*

$$\lambda x = \lambda u_i(x_i) = u_i(\lambda x_i).$$

(d): *Consequently, if the maps* u_{ji} *are injective, the space* E_i *can be identified with a vector subspace of* E *via the canonical injection* $u_i:E_i \to E$.

(e): *Finally, it can be shown (although we shall not use it) that the vector space E is a quotient of the vector space direct sum of the E_i's.*

For a detailed exposition of the theory of inductive limits see Bourbaki [*1*,*2*].

0•A.2 PROJECTIVE LIMITS OF VECTOR SPACES

Here I is as in Section 0•A.2.

0•A.2˙1 Projective Systems of Vector Spaces

Let $(E_i)_{i \in I}$ be a family of vector spaces over $\mathbb{K}$. Suppose that for every pair $(i,j) \in I \times I$ such that $i \leqslant j$, there exists a linear map $p_{ij}: E_j \to E_i$ such that the system of maps (p_{ij}) satisfies the following conditions:

(i) For every $i \in I$, p_{ii} is the identity of E_i;

(ii) For every $i \leqslant j \leqslant k$, $p_{ik} = p_{ij} \circ p_{jk}$.

Then the system (E_i, p_{ij}) is called a *PROJECTIVE SYSTEM OF VECTOR SPACES over* $\mathbb{K}$.

0•A.2˙2 Existence and Uniqueness of the Projective Limit of Vector Spaces

Let (E_i, p_{ij}) be a projective system of vector spaces over $\mathbb{K}$. There exist a vector space E over $\mathbb{K}$ and, for each $i \in I$, a linear map $p_i: E \to E_i$, such that:

(PL.1): $p_i = p_{ij} \circ p_j$ *whenever* $i \leqslant j$;

(PL.2): *For every vector space F and family of linear maps $q_i: F \to E_i$ such that $q_i = p_{ij} \circ q_j$ for $i \leqslant j$, there exists a* unique *linear map $q: F \to E$ satisfying $q_i = p_i \circ q$ for all $i \in I$.*

The vector space E is unique up to isomorphism and is called the *PROJECTIVE LIMIT* of the projective system (E_i, p_{ij}). For every $i \in I$, the map $p_i: E \to E_i$ is called the *CANONICAL PROJECTION* of E onto E_i.

0•A.2˙3 Properties of Projective Limits

From the set-theoretical and algebraic points of view E is a vector subspace of the product $\prod_{i \in I} E_i$. This subspace is defined as the set of all elements $x = (x_i)_{i \in I}$ of $\prod_{i \in I} E_i$ with $p_{ij}(x_j) = x_i$ for all $i \leqslant j$. For each $i \in I$, $p_i: E \to E_i$ is the restriction to E of the canonical projection of $\prod_{i \in I} E_i$ onto E_i.

A detailed study of projective limits can be found in Bourbaki [*1*,*2*].

0·A.3 DISKS IN VECTOR SPACES

0·A.3˙1 Notation

Let E be a *vector space* over $\mathbb{K}$. For *SUBSETS* A and B of E and Λ of $\mathbb{K}$ we write:

$$A + B = \{x \in E;\ x = a + b \text{ with } a \in A \text{ and } b \in B\},$$

$$\Lambda A = \{x \in E;\ x = \lambda a \text{ with } \lambda \in \Lambda \text{ and } a \in A\}.$$

If A consists of a single point x, we write $x + B$ for $\{x\} + B$ and Λx for $\Lambda\{x\}$. Similarly, when Λ consists of a single scalar λ, we write λA for $\{\lambda\}A$.

$A + B$ is the *VECTOR SUM of* A *and* B, and, for every $\lambda \in \mathbb{K}$, $\lambda \neq 0$, λA is the *HOMOTHETIC IMAGE* of A under the *HOMOTHETIC TRANSFORMATION* $A \to \lambda A$.

In the following $\lambda, \mu, \alpha, \ldots$ denote scalars.

0·A.3˙2 Circled, Convex and Absorbent Sets: Disks

0·A.3˙2(a) *Definitions*

Let A and B be two subsets of a vector space E. We say that:

(i): *A is CIRCLED if $\lambda A \subset A$ whenever $\lambda \in \mathbb{K}$ and $|\lambda| \leqslant 1$;*

(ii): *A is CONVEX if $\lambda A + \mu A \subset A$ whenever λ and μ are positive real numbers such that $\lambda + \mu = 1$;*

(iii): *A is DISKED, or a DISK, if A is both convex and circled;*

(iv): *A ABSORBS B if there exists $\alpha \in \mathbb{R}$, $\alpha > 0$, such that $\lambda A \supset B$ whenever $|\lambda| \geqslant \alpha$;*

(v): *A is ABSORBENT in E if A absorbs every subset of E* consisting of a single point.

0·A.3˙2(b) *Elementary Properties*

Let A and B be as in Subsection 0 A.3˙2(a).

(i): *If A is circled, $\lambda A = |\lambda| A$ and if $|\lambda| \leqslant |\mu|$, $\lambda A \subset \mu A$.*

(ii): *If A is circled, then A absorbs B if there exists $\lambda > 0$ such that $B \subset \lambda A$.*

(iii): *If A and B are convex and $\lambda, \mu \in \mathbb{K}$, then $\lambda A + \mu B$ is convex.*

(iv): *Every intersection of circled (resp. convex) sets is circled (resp. convex); hence an intersection of disks is again a disk.*

(v): *Let E and F be vector spaces and let $u: E \to F$ be a linear map. Then the image, direct or inverse, under u of a circled (resp. convex) subset is circled (resp. convex).*

(vi): *A subset A is a disk if and only if $\lambda A + \mu A \subset A$ whenever λ and μ are scalars such that $|\lambda| + |\mu| \leqslant 1$.*

All these assertions are evident. As an example, we shall verify (vi). Let A be a disk and let λ, μ be non-zero scalars such that $|\lambda| + |\mu| \leqslant 1$. Then for $x \in A$ and $y \in A$ we have:

$$\lambda x + \mu y = (|\lambda| + |\mu|)\left[\frac{|\lambda|}{|\lambda| + |\mu|}\left(\frac{\lambda}{|\lambda|} x\right) + \frac{|\mu|}{|\lambda| + |\mu|}\left(\frac{\mu}{|\mu|} y\right)\right].$$

Since A is circled $\lambda x/|\lambda|$ and $\mu y/|\mu|$ belong to A, hence so does the term in square brackets, for A is convex. But then $\lambda x + \mu y \in A$, since $|\lambda| + |\mu| \leqslant 1$ and A is circled. The converse is obvious.

0·A.3˙3 Circled, Convex and Disked Hulls

0·A.3˙3(a) *Notations and Definitions*

Since the intersection of circled (resp. convex, disked) subsets is circled (resp. convex, disked) and the whole space E is disked, for every subset $A \subset E$ there exists a smallest circled (resp. convex, disked) subset containing A. This subset is called the *CIRCLED* (resp. *CONVEX*, *DISKED*) *HULL of A*.

It is easily seen that the circled hull of A is the set $\bigcup_{|\lambda| \leqslant 1} \lambda A$. We shall *denote by* co(A) (resp. $\Gamma(A)$) *the convex (resp. disked) hull of A.*

0·A.3˙3(b) *Characterisation of the Convex Hull*

PROPOSITION (1): *Let* $(A_i)_{i \in I}$ *be an arbitrary family of convex subsets of a vector space E. The convex hull of* $\bigcup_{i \in I} A_i$ *is the set C of all linear combinations of the form* $\sum_{i \in I} \lambda_i x_i$, *where* $x_i \in A_i$, $\lambda_i > 0$, $\sum_{i \in I} \lambda_i = 1$ *and only finitely many* λ_i*'s are non-zero.*

Proof: Clearly $C \supset \bigcup_{i \in I} A_i$. We show that C is convex. Let $x = \sum_{i \in I} \lambda_i x_i$ and $y = \sum_{i \in I} \mu_i y_i$ be two elements of C and let α, β be non-negative scalars such that $\alpha + \beta = 1$. We have to show that $\alpha x + \beta y \in C$. This is obvious if either $\alpha = 0$ or $\beta = 0$; hence we may assume α and β to be positive. For each $i \in I$ let $\nu_i = \alpha\lambda_i + \beta\mu_i$ and denote by J the finite subset of I such that $\nu_i > 0$ for $i \in J$. If:

$$z_i = \frac{\alpha\lambda_i x_i + \beta\mu_i y_i}{\alpha\lambda_i + \beta\mu_i} \qquad \text{for all} \quad i \in J,$$

then $z_i \in A_i$ and $\alpha x + \beta y = \sum_{i \in I} \nu_i z_i \in C$ since $\sum_{i \in I} \nu_i = 1$. Consequently C is convex and it remains to show that C is contained in every convex set containing $\bigcup_{i \in I} A_i$. Let B be such a convex set and let $x \in C$; then x can be written as $x = \sum_{i=1}^{n} \lambda_i x_i$, with $\sum_{i=1}^{n} \lambda_i = 1$, $\lambda_i \geqslant 0$, $x_i \in A \subset B$ and n a positive integer. If $n = 1$, then clearly $x \in B$, since B is convex. Inductively, we assume that $y \in B$ whenever $y = \sum_{i=1}^{n} \mu_i x_i$ and $k \leqslant n - 1$. We may also assume all λ_i's to be positive, $i = 1,\ldots,n$. Let $\alpha = \sum_{i=1}^{n-1} \lambda_i$, $\beta = \lambda_n$ and $\mu_i = \lambda_i/\alpha$, $i = 1,\ldots,n-1$. The induction hypothesis then ensures that $\sum_{i=1}^{n-1} \mu_i x_i \in B$, hence, by definition of a convex set, $\sum_{i=1}^{n} \lambda_i x_i = \alpha\left(\sum_{i=1}^{n-1} \mu_i x_i\right) + \beta x_n \in B$ and the proposition is completely proved.

COROLLARY (1): *Let A be a subset of E. The convex hull of A in E is the set of all finite linear combinations of the form $\sum_{i \in I} \lambda_i x_i$, where $\lambda_i \geqslant 0$, $\sum_{i \in I} \lambda_i = 1$ and $x_i \in A$.*

In fact $A = \bigcup_{x \in A} \{x\}$.

COROLLARY (2): *The convex hull of a circled set is circled, hence disked.*

Indeed, let A be circled and let $B = \mathrm{co}(A)$. If $x \in B$, then $x = \sum \lambda_i x_i$ with $\sum \lambda_i = 1$ and $\lambda_i \geqslant 0$. Let $\alpha \in \mathbb{C}$ with $|\alpha| \leqslant 1$; since A is circled, $\alpha x_i \in A \subset B$, whence $\alpha x = \sum \alpha\lambda_i x_i = \sum \lambda_i(\alpha x_i) \in B$.

0·A.3·3(c) *Characterisation of the Disked Hull*

PROPOSITION (2): *The disked hull of a subset $A \subset E$ is the convex hull of the circled hull of A.*

Proof: Let B be the convex hull of the circled hull of A. B is a disk (Proposition (1), Corollary (2)) containing A, hence also $\Gamma(A)$. Conversely, $\Gamma(A)$ is disked, hence circled, and contains A, whence it contains the circled hull of A. Since $\Gamma(A)$ is also convex, it contains the convex hull of the circled hull of A, i.e. B.

PROPOSITION (3): *Let A be a subset of E. The disked hull $\Gamma(A)$ of A is the set of finite linear combinations of the form $\sum_{i \in I} \lambda_i x_i$, with $x_i \in A$, $\lambda_i \in \mathbb{K}$ and $\sum_{i \in I} |\lambda_i| \leqslant 1$.*

Proof: Let:

$$D = \{x = \sum \lambda_i x_i;\ x_i \in A,\ \lambda_i \in \mathbb{K},\ \sum |\lambda_i| \leqslant 1\},$$

and denote by B the circled hull of A. By Proposition (2) $\Gamma(A) =$

co(B) and so:

$$\Gamma(A) = \{x = \sum \alpha_i y_i;\ y_i \in B,\ \alpha_i \geq 0,\ \sum \alpha_i = 1\}.$$

Let then $x \in \Gamma(A)$ be of the form $\sum \alpha_i y_i$, $y_i \in B$. Since $B = \bigcup_{|\lambda| \leq 1} \lambda A$ there exist scalars λ_i ($|\lambda_i| \leq 1$) such that $y_i = \lambda_i x_i$, with $x_i \in A$. Thus $x = \sum \alpha_i \lambda_i x_i$ and $\sum |\alpha_i \lambda_i| = \sum \alpha_i |\lambda_i| \leq \sum \alpha_i = 1$, implying that $x \in D$ and, consequently, that $\Gamma(A) \subset D$. Conversely, if $x \in D$, then $x = \sum_{i=1}^{n} \lambda_i x_i$, with $x_i \in A$, $\sum_{i=1}^{n} |\lambda_i| \leq 1$ and $n \in \mathbb{N}$ (the positive integers). In order to show that $x \in \Gamma(A)$ it suffices to prove that:

$$\sum_{i=1}^{n} (\lambda_i A) \subset \left(\sum_{i=1}^{n} |\lambda_i|\right)\Gamma(A). \qquad (*)$$

This is clear if $n = 1$, since $\Gamma(A)$ is circled. For $n = 2$ we show that $\lambda A + \mu A \subset (|\lambda| + |\mu|)\Gamma(A)$. This is evident if $\lambda = \mu = 0$ and hence we may assume $|\lambda| + |\mu| \neq 0$. Since $\Gamma(A)$ is disked:

$$\frac{\lambda}{|\lambda| + |\mu|} A + \frac{\mu}{|\lambda| + |\mu|} A \subset \Gamma(A).$$

giving $\lambda A + \mu A \subset (|\lambda| + |\mu|)\Gamma(A)$. Finally, supposing (*) to be true for $1, 2, \ldots, n - 1$, we have:

$$\begin{aligned}\sum_{i=1}^{n} (\lambda_i A) &= \sum_{i=1}^{n=1} (\lambda_i A) + \lambda_n A \\ &\subset \left(\sum_{i=1}^{n} |\lambda_i|\right)\Gamma(A) + |\lambda_n|\Gamma(A) \\ &= \left(\sum_{i=1}^{n} |\lambda_i|\right)\Gamma(A).\end{aligned}$$

which is (*).

0·A.4 GAUGES OF DISKS AND SEMI-NORMS

Let $\mathbb{R}_+$ be the set of non-negative real numbers and let E be a vector space over $\mathbb{K}$.

DEFINITION (1): *A SEMI-NORM on E is a map $p: E \to \mathbb{R}_+$ with the following properties:*

(i) $p(\lambda x) = |\lambda| p(x)$ *for every* $\lambda \in \mathbb{K}$;

(ii) $p(x + y) \leq p(x) + p(y)$.

Clearly $p(0) = p(0.x) = 0p(x) = 0$, but it is possible that $p(x) = 0$ with $x \neq 0$. p is called a *NORM* if $p(x) = 0$ implies $x = 0$. A *SEMI-NORMED VECTOR SPACE* (resp. *NORMED VECTOR SPACE*) is a pair (E,p) consisting of a vector space E and a semi-norm (resp. norm) p on E. The sets $A = \{x \in E;\ p(x) < 1\}$ and $B = \{x \in E;\ p(x) \leqslant 1\}$ are called respectively the *OPEN* and *CLOSED UNIT BALL* of (E,p). The unit ball (open or closed) of a semi-normed vector space is an absorbent disk in E. Conversely, with every absorbent disk of a vector space E we can associate a semi-norm on E as follows:

DEFINITION (2): *Let A be an absorbent disk in a vector space E. The GAUGE of A, denoted by p_A, is the map of E into $\mathbb{R}_+$ defined by:*

$$p_A(x) = \inf\{\alpha \in \mathbb{R}_+;\ x \in \alpha A\}.$$

PROPOSITION (1): *The gauge of an absorbent disk in E is a semi-norm on E.*

Proof: Since A is absorbent in E, p_A takes its values in $\mathbb{R}_+$. Let $x \in E$ and $\lambda \in \mathbb{K}$; if $\lambda = 0$, it is clear that $p_A(\lambda x) = |\lambda| p(x) = 0$. Otherwise:

$$p_A(\lambda x) = \inf\{\alpha > 0;\ \lambda x \in \alpha A\} = \inf\{\alpha > 0;\ x \in \frac{\alpha}{|\lambda|} A\},$$

since A is circled; also:

$$|\lambda| p_A(x) = |\lambda| \inf\{\beta > 0;\ x \in \beta A\} = \inf\{|\lambda|\beta > 0;\ x \in \beta A\}$$

$$= \inf\{\alpha = |\lambda|\beta > 0;\ x \in \frac{\alpha}{|\lambda|} A\},$$

and consequently $p_A(\lambda x) = |\lambda| p_A(x)$.

To verify the 'triangle inequality' consider $x, y \in E$. Since A is absorbent, there exist $\alpha, \beta > 0$ such that $x \in \alpha A$ and $y \in \beta A$. It follows that $x + y \in \alpha A + \beta A = (\alpha + \beta)A$, i.e. $p_A(x + y) \leqslant \alpha + \beta$, and, finally, $p_A(x + y) \leqslant p_A(x) + p_A(y)$, since α and β are arbitrary.

REMARK (1): Let (E,p) be a semi-normed vector space and let A be the open unit ball (resp. let B be the closed unit ball) of (E,p). Then $p_A = p_B = p$.

REMARK (2): If A is an absorbent disk in E, then:

$$\{x \in E;\ p_A(x) < 1\} \subset A \subset \{x \in E;\ p_A(x) \leqslant 1\}.$$

0·A.5 THE SPACES E_A

0·A.5·1 Definitions

Let E be a vector space and let A be a disk in E *not necessarily absorbent* in E. We denote by E_A the vector space spanned by

A, i.e. the space $\bigcup_{\lambda>0} \lambda A = \bigcup_{\lambda\in\mathbb{K}} \lambda A$. The disk A is absorbent in E_A and we can then endow E_A with the semi-norm p_A, gauge of A. This semi-norm p_A is called the *CANONICAL SEMI-NORM* of E_A and A is said to be *NORMING* if p_A is a norm on E_A.

0·A.5·2 The Space $E_{(A+B)}$

Let A and B be two disks in a vector space E. Clearly $\Gamma(A\cup B) \subset A + B \subset 2\Gamma(A\cup B)$. Consequently $E_{(A+B)} = E_{\Gamma(A\cup B)}$ and on this vector space the semi-norms p_{A+B} and $p_{\Gamma(A\cup B)}$ are equivalent. The semi-norm p_{A+B} can be expressed in terms of the semi-norms p_A and p_B. Precisely, we have:

LEMMA (1): (i) $E_{(A+B)} = E_A + E_B$;

(ii) $$p_{A+B}(x) = \inf_{\substack{x=y+z \\ y\in E_A, z\in E_B}} \max(p_A(y), p_B(z)), \qquad x\in E_{A+B}.$$

Proof: (i): $E(A+B) \subset E_A + E_B$, since $A + B \subset E_A + E_B$. On the other hand, since $A\cup B \subset A + B$, E_A and E_B are contained in $E_{(A+B)}$ and hence:

$$E_A + E_B \subset E_{(A+B)} + E_{(A+B)} = E_{(A+B)}.$$

(ii): Denote by λ the right hand side and let $x\in E_{(A+B)} = E_A + E_B$ be of the form $x = y + z$, $y\in E_A$ and $z\in E_B$. Let $\alpha = \max(p_A(y), p_B(z))$. Then for $\varepsilon > 0$, $y\in(\alpha+\varepsilon)A$ and $z\in(\alpha+\varepsilon)B$, hence $x\in(\alpha+\varepsilon)(A + B)$, i.e. $p_{A+B}(x) \leqslant \alpha + \varepsilon$, whence, since ε is arbitrary, $p_{(A+B)}(x) \leqslant \lambda$. Conversely, we show that $\lambda \leqslant p_{A+B}(x)$. Suppose not; then $\lambda > p_{A+B}(x) = \inf\{\alpha > 0;\ x\in\alpha(A + B)\}$ and there exists $\alpha > 0$ such that $x\in\alpha(A + B)$ and $\alpha < \lambda$. It follows that $x\in\alpha A + \beta B$ and hence $x = y + z$ with $y\in\alpha A \subset E_A$ and $z\in\alpha B\subset E_B$. Thus $p_A(y) \leqslant \alpha$, $p_B(z) \leqslant \alpha$ and consequently $\lambda \leqslant \alpha$, contradicting the choice of α. This proves the Lemma.

Before stating an important consequence of this Lemma, let us recall some definitions. If (Y,p) and (Z,q) are two semi-normed spaces, a semi-norm can be defined on the vector space product $Y\times Z$ by setting:

$$r(y,z) = \max(p(y), q(z)), \qquad y\in Y,\ z\in Z,$$

Such a semi-norm on $Y\times Z$ is called the *PRODUCT SEMI-NORM*, and the vector space $Y\times Z$, when furnished with the product semi-norm, is called the *SEMI-NORMED PRODUCT SPACE* of the semi-normed spaces (Y,p) and (Z,q).

Let now M be a vector subspace of (Y,p). We define a semi-norm $\dot{p}$ on the vector space quotient Y/M by setting:

$$\dot{p}(\dot{x}) = \inf\{p(t);\ t\in\dot{x}\}, \qquad \dot{x}\in Y/M.$$

The semi-norm $\dot{p}$ is called the *QUOTIENT SEMI-NORM* of p on Y/M and

$(Y/M,\dot{p})$ is called the *SEMI-NORMED QUOTIENT SPACE* of Y by M.

PROPOSITION (1): *Let E be a vector space and let A,B be two disks in E.*

(i) *The map $u: E_A \times E_B \to E_{A+B}$ defined by $u(y,z) = y + z$ is a linear surjection, whose kernel we denote by M.*

(ii) *If we endow $E_A \times E_B$ with the product semi-norm $p(y,z) = \max(p_A(y), p_B(z))$ and $(E_A \times E_B)/M$ with the quotient semi-norm of p, then the semi-normed spaces $E_{(A+B)}$ and $(E_A \times E_B)/M$ are isometric.*

Proof: The map u is evidently linear, and a surjection since $E_A + E_B = E_{A+B}$. Let M be its kernel. The quotient semi-norm on $(E_A \times E_B)/M$ is given by:

$$\dot{p}(\dot{x}) = \inf\{p(y,z);\ (y,z) \in \dot{x},\ y \in E_A,\ z \in E_B\},$$

for every $x \in E_A \times E_B$. Thus it suffices to verify that the quotient map:

$$\dot{u}: (E_A \times E_B)/M \to E_{(A+B)}$$

is an isometry, i.e. that $p_{A+B}(u(x)) = \dot{p}(\dot{x})$ for every $x \in E_A \times E_B$. For the left hand side we have, by Lemma (1):

$$p_{A+B}(u(x)) = \inf_{\substack{u(x)=a+b \\ a\in E_A,\, b\in E_B}} \max(p_A(a), p_B(b)),$$

and for the right hand side:

$$\dot{p}(\dot{x}) = \inf_{\substack{(a,b)\in\dot{x} \\ a\in E_A,\, b\in E_B}} \max(p_A(a), p_B(b)).$$

Therefore, $p_{A+B}(u(x)) = \dot{p}(\dot{x})$, since $(a,b) \in \dot{x}$ is equivalent to $u(a,b) = u(x)$.

0·A.5˙3 Infinite Products of Disks

The following Proposition is an immediate consequence of the definitions:

PROPOSITION (2): *Let $(E_i)_{i\in I}$ be a family of vector spaces indexed by a non-empty set I and, for every $i \in I$, let A_i be a disk in E_i with gauge $\|\cdot\|_i$. Let $E = \prod_{i\in I} E_i$ be the vector space product of the spaces E_i and set $A = \prod_{i\in I} A_i$. Then A is a disk and, moreover:*

(i) $E_A = \{x = (x_i) : \sup_{i\in I} \|x_i\|_i < +\infty\}$;

(ii) $p_A(x) = \sup_{i\in I} \|x_i\|_i$, *where p_A is the gauge of A.*

0·A.5·4 Effect of a Linear Map

PROPOSITION (3): *Let E,F be two vector spaces and $u:E \to F$ a linear map. For every disk A in E, $u(A)$ is a disk in F and $F_{u(A)}$ is isometric to the semi-normed space E_A/N, where $N = (\ker u) \cap E_A$*

Proof: The restriction of u to E_A is a surjection onto $F_{u(A)}$ with kernel N, hence an algebraic isomorphism between $F_{u(A)}$ and E_A/N. Moreover, for every $y \in F_{u(A)}$ we have:

$$p_{u(A)}(y) = \inf_{\substack{\lambda>0 \\ y\in\lambda u(A)}} \lambda = \inf_{y=u(x)} \left(\inf_{x\in\lambda A} \lambda \right) = \inf_{y=u(x)} p_A(x),$$

and the assertion follows.

0·B PRELIMINARIES OF GENERAL TOPOLOGY AND NORMED SPACES

We assume the reader to be familiar with the fundamental notions of *General Topology* and of the *theory of normed vector spaces*. Specifically, the reader should have some knowledge of:

— The definition of a topological space by the axioms of open sets, closed sets and neighbourhoods;

— The following notions: separated space, closure, topological subspace, product of topological spaces, quotient of of a topological space, continuous map;

— The all important notion of *compact set*, key notion of all modern Analysis, together with Tychonov's Theorem: 'Every product of compact sets is compact'. This Theorem will only be used in Chapter VI;

— Ascoli's Theorem, used in Chapter VIII;

— Metric spaces, metrizable spaces, completeness and Baire's Theorem, of which we shall only use the following particular case: 'Every Banach space is a Baire space' (Chapter IV).

— The most elementary notions on normed and Banach spaces including F. Riesz's Finiteness Theorem: 'A Banach space is finite-dimensional if and only if its unit ball is compact'. Hilbert spaces, however, will not be used.

All the abovementioned notions and results can be found in the classical introductory texts on Topology and Functional Analysis, for example G. Choquet [1], J. Dieudonné [1,2] and L. Schwartz [1], to which the reader is referred.

0·C TOPOLOGICAL VECTOR SPACES

0·C.1 DEFINITION

Let E be a vector space over $\mathbb{K}$ ($=\mathbb{R}$ or $\mathbb{C}$). A topology on E is said to be a *VECTOR TOPOLOGY* or a *TOPOLOGY COMPATIBLE WITH THE VECTOR SPACE STRUCTURE of* E if the following maps are continuous when E is endowed with such a topology:

(i) $(x,y) \to x + y$ from $E \times E$ into E;

(ii) $(\lambda,x) \to \lambda x$ from $\mathbb{K} \times E$ into E.

Here $E \times E$ and $\mathbb{K} \times E$ are assumed to have their product topologies, $\mathbb{K}$ being given the topology defined by taking its absolute value as a norm.

We call *TOPOLOGICAL VECTOR SPACE* any vector space E endowed with a topology compatible with the vector space structure of E. For a topological vector space the system of neighbourhoods of a point x is completely determined by the system of neighbourhoods of 0. Precisely, *if $\mathcal{V}$ is a base of neighbourhoods of* 0 *in a topological vector space E, then for every point $x \in E$ the family:*

$$\mathcal{V}(x) = \{x + V\colon V \in \mathcal{V}\}$$

is a base of neighbourhoods of x in E. In fact, if $V \in \mathcal{V}$, then $x + V$ in a neighbourhood of x, being the inverse image of V under the map $y \to y - x$ which is continuous from E into E by (i). Conversely, if U is an arbitrary neighbourhood of x, then $U = V + x$, where V is the inverse image of U under the continuous map $y \to y + x$ hence $V \in \mathcal{V}$.

The above property is expressed by saying that the topology of a topological vector space is a *TRANSLATION-INVARIANT TOPOLOGY*. As a consequence of it we may, and shall, only concentrate upon the system of neighbourhoods of zero.

0·C.2 CHARACTERISATION OF THE FILTER OF NEIGHBOURHOODS OF ZERO

0·C.2˙1 Notion of Filter and Filter Base of a Set

Let X be a set. A non-empty family $\mathfrak{F}$ of subsets of X is a *FILTER on* X if $\mathfrak{F}$ satisfies the following three axioms:

(i) *The empty set does not belong to* $\mathfrak{F}$;

(ii) *A finite intersection of elements of* $\mathfrak{F}$ *is an element of* $\mathfrak{F}$;

(iii) *Every subset A of X which contains an element of* $\mathfrak{F}$ *belongs to* $\mathfrak{F}$.

A *FILTER BASE on* X is any non-empty family $\mathcal{B}$ of subsets of X

satisfying the following two axioms:

(i) *No element of* $\mathcal{B}$ *is empty;*

(ii) *The intersection of any two elements of* $\mathcal{B}$ *contains an element of* $\mathcal{B}$.

If $\mathcal{B}$ is a filter base on X, the family $\mathcal{F}$ of subsets of X which contain at least one element of $\mathcal{B}$ is a filter on X called the *FILTER GENERATED BY* $\mathcal{B}$. A fundamental example of a filter is exhibited by the family of neighbourhoods of a point x in a topological space X. Any base of neighbourhoods of x is then a base for the filter of neighbourhoods of x.

0·C.2·2

THEOREM (1): (a) *Let E be a topological vector space. There exists a base* $\mathcal{B}$ *of neighbourhoods of zero in E consisting of closed sets such that:*

(i) *Each* $V \in \mathcal{B}$ *is absorbent and circled;*

(ii) *For every* $V \in \mathcal{B}$ *and* $\lambda \neq 0$ *in* $\mathbb{K}$, $\lambda V \in \mathcal{B}$ *(invariance under homothetic transformations);*

(iii) *For every* $V \in \mathcal{B}$ *there exists* $W \in \mathcal{B}$ *such that* $W + W \subset V$.

(b) *Conversely, let E be a vector space over* $\mathbb{K}$ *and let* $\mathcal{B}$ *be a filter base on E satisfying* (i,ii,iii). *Then there exists one and only one topology on E, compatible with the vector space structure of E, for which* $\mathcal{B}$ *is a base of neighbourhoods of* 0.

The proof of this Theorem can be found in the literature, e.g. L. Schwartz [*1*].

0·C.3 On the Closure of Disks in Topological Vector Spaces

PROPOSITION (1): *In a topological vector space the closure of a circled (resp. convex) set is again circled (resp. convex).*

Proof: (i): Let A be a circled subset of a topological vector space E and let D be the closed unit ball of $\mathbb{K}$. Denote by u the map $(\lambda,x) \to \lambda x$ of $\mathbb{K} \times E$ into E. Since A is circled, $u(D \times A) \subset A$ and since u is continuous, $u(D \times \bar{A}) = u(\overline{D \times A}) \subset \overline{u(D \times A)} \subset \bar{A}$, which shows that $\bar{A}$ is circled.

(ii): Let A be a convex subset of E. For every $\lambda \in [0,1]$ we denote by u_λ the continuous map $(x,y) \to \lambda x + (1 - \lambda)y$ of $E \times E$ into E. The convexity of A implies that $u_\lambda(A \times A) \subset A$, hence, since u_λ is continuous, $u_\lambda(\bar{A} \times \bar{A}) = u_\lambda(\overline{A \times A}) \subset \bar{A}$. Thus $\bar{A}$ is convex.

COROLLARY: *In a topological vector space the closure of a disk A is again a disk. Indeed, it is the smallest closed disk containing A.*

The smallest closed disk containing a subset A of E is called the *CLOSED DISKED HULL* of A and is clearly the closure of the disked hull of A.

0·C.4 LOCALLY CONVEX SPACES

0·C.4˙1 Definition and Characterisation of the Filter of Neighbourhoods of Zero

DEFINITION (1): *A topological vector space possessing a base of neighbourhoods of* 0 *which consists of convex sets is called a LOCALLY CONVEX SPACE.*

PROPOSITION (1): *Every locally convex space has a base of neighbourhoods of* 0 *consisting of closed disks.*

Proof: Let E be a locally convex space. As a topological vector space E possesses a base of closed neighbourhoods of 0 (Theorem (1) of Subsection 0·C.2˙2). Let V be one such neighbourhood; V contains a convex neighbourhood U of 0. Now the closure of U is again convex (Section 0·C.3), whence V contains the closed convex neighbourhood $\bar{U}$ of 0. This shows that E has a base of closed convex neighbourhoods of 0 and it remains to prove that any such neighbourhood contains a closed disked one. Let W be a closed convex neighbourhood of 0 and let $N = \bigcap_{|\mu|\geqslant 1} \mu W$. The set N is closed, convex and circled, hence is a closed disk, clearly contained in W. It is then enough to show that N is a neighbourhood of 0 in E. By the continuity of the map $(\lambda,x) \to \lambda x$ of $\mathbb{K}\times E$ into E at the point $(0,0)$, there exists a real number $\alpha > 0$ and a neighbourhood M of 0 such that $\bigcup_{|\lambda|\leqslant\alpha} \lambda M \subset W$. Clearly $\alpha M \subset N$; in fact, if $\mu \in \mathbb{K}$ and $|\mu| > 1$, then $\alpha/|\mu| \leqslant \alpha$, hence $\lambda M \subset W$ where $\lambda = \alpha/\mu$, i.e. $\alpha M \subset \mu W$. Since μ is arbitrary, we must have $\alpha M \subset N$, which completes the proof.

Conversely, we have:

PROPOSITION (2): *Let E be a vector space and let $\mathcal{B}$ be a filter base on E consisting of absorbent disks and invariant under homothetic transformations. Then $\mathcal{B}$ is a base of neighbourhoods of* 0 *for a locally convex topology on E.*

Proof: By part (b) of Theorem (1) of Subsection 0·C.2˙2 it suffices to show that $\mathcal{B}$ satisfies condition (iii) of that Theorem However, this is clear, since, every subset $V \in \mathcal{B}$ being convex, we have $W + W \subset V$ if $W = \frac{1}{2}V$; also $W \in \mathcal{B}$ (invariance under homothetic transformations), hence $\mathcal{B}$ defines a vector topology, obviously locally convex, on E.

0·C.4˙2 Pro-Normed Character of Locally Convex Spaces

Let E be a locally convex space and $\mathcal{V}$ a base of neighbourhoods of 0 in E consisting of disks (necessarily absorbent). For every

$V \in \mathcal{V}$ the vector space spanned by V is E and we denote by (E,V) the vector space E endowed with the semi-norm p_V gauge of V.

The important fact here is that *the topology of E is completely determined by that of the semi-normed spaces (E,V).* Precisely, let us *denote by $\varphi_V : E \to (E,V)$ the identity of E.* Then:

> PROPOSITION (3): *The topology of E is the coarsest topology on E which makes the maps φ_V continuous.*

Proof: Since V is a neighbourhood of 0 in E, φ_V is clearly continuous when E is given its initial topology $\mathcal{T}$. Let $\mathcal{T}'$ be another topology on E for which the maps φ_V are continuous. We have to show that the identity $(E,\mathcal{T}') \to (E,\mathcal{T})$ is continuous. But this is evident, for if U is a disked neighbourhood of 0 in $(E,\mathcal{T})$ then U is a neighbourhood of 0 in (E,U) and hence $\varphi_U^{-1}(U)$ is a neighbourhood of 0 in $(E,\mathcal{T}')$.

The above Proposition is usually stated by saying that every locally convex topology is an initial topology for a family of semi-normed topologies. *The family of semi-norms $\{p_V : V \in \mathcal{V}(0)\}$ is called the FAMILY OF SEMI-NORMS ASSOCIATED WITH THE TOPOLOGY of E.*

0·C.4·3 Topologies Defined by a Family of Semi-Norms

Let E be a vector space and $\Gamma = (p_i)_{i \in I}$ a family of semi-norms on E. For every $i \in I$ let $E_i = (E, p_i)$ be the vector space E furnished with the semi-norm p_i and let $\varphi_i : E \to E_i$ be the identity of E. The coarsest topology on E for which each of the maps φ_i is continuous is a locally convex topology. In fact, such a topology has a base of neighbourhoods of 0 given by all intersections $\bigcap_{i \in J} \varphi_i^{-1}(V_i)$ where V_i is any neighbourhood of 0 in E_i and J is any finite subset of I. *This topology is said to be the TOPOLOGY GENERATED BY THE FAMILY Γ OF SEMI-NORMS ON E.*

Proposition (3) asserts that every locally convex topology is generated by a family of semi-norms. Since, conversely, every topology defined by a family of semi-norms is locally convex, we have the *equivalence between the notion of a locally convex topology and that of a topology generated by a family of semi-norms.*

0·C.4·4 Convergence in a Locally Convex Space

Let X be an arbitrary set. A *NET in X* is any map of a directed set Λ into X and is denoted by $(x_\lambda)_{\lambda \in \Lambda}$. If $\Lambda = \mathbb{N}$ we recover the usual notion of a *SEQUENCE* $(x_n)_{n \in \mathbb{N}}$.

Suppose now that X is a topological space. A net $(x_\lambda)_{\lambda \in \Lambda}$ in X is said to *CONVERGE* to a point $x \in X$ if the following condition is satisfied:

> *For every neighbourhood V of x, there exists $\lambda_0 \in \Lambda$ such that $x_\lambda \in W$ whenever $\lambda > \lambda_0$, where $>$ is the order relation on Λ.*

As an immediate consequence of this definition we have:

PROPOSITION (4): *Suppose that E is a locally convex space and that $(p_i)_{i\in I}$ is a family of semi-norms generating the topology of E. A net $(x_\lambda)_{\lambda\in\Lambda}$ in E converges to 0 if and only if, for each $i\in I$, $(p_i(x_\lambda))_{\lambda\in\Lambda}$ converges to 0 in $\mathbb{R}$.*

0·C.4·5 Metrizable Topological Vector Spaces

A topological vector space is said to be *METRIZABLE* if it has a *countable* base of neighbourhoods of the origin. If $U_1, U_2, \ldots, U_n, \ldots$ is such a base, setting $V_n = U_1 \cap U_2 \cap \ldots \cap U_n$, we obtain a new countable base of neighbourhoods of 0 which is decreasing, in the sense that $V_n \supset V_{n+1}$ for every $n \in \mathbb{N}$.

It can be shown that metrizable topological vector spaces are exactly those topological vector spaces whose topology is metrizable, i.e. that may be defined in terms of a distance function. The locally convex space case deserves special attention.

PROPOSITION (5): *A locally convex space is metrizable if and only if it is separated and its topology is defined by a countable family of semi-norms.*

Proof: Let E be a locally convex space. If E is metrizable its origin has a countable base (V_n) of disked neighbourhoods. Denoting by p_n the gauge of V_n, it is clear that the sequence of semi-norms (p_n) defines the topology of E. Conversely, if (p_n) is a sequence of semi-norms generating the topology of E, a countable base of neighbourhoods of 0 is exhibited by the sets:

$$V_n = \{x \in E;\ p_n(x) \leqslant 1\}.$$

0·C.4·6 Complete Topological Vector Spaces

Let E be a separated topological vector space. A net $(x_\lambda)_{\lambda\in\Lambda}$ in E is called a *CAUCHY NET* if for every neighbourhood V of 0 in E there exists $\lambda_0 \in \Lambda$ such that $x_\lambda - x_{\lambda'} \in V$ whenever $\lambda > \lambda_0$ and $\lambda' > \lambda_0$. For $\Lambda = \mathbb{N}$ we obtain the familiar notion of a *CAUCHY SEQUENCE*.

A subset A of E is said to be *COMPLETE* (resp. *SEQUENTIALLY COMPLETE*) if every Cauchy net (resp. Cauchy sequence) in A converges to an element of A. The notion of completeness (resp. sequential completeness) for E is obtained by taking $A = E$ in the above definition.

An equivalent defintion of completeness in terms of Cauchy filters can be found in the literature.

It is clear that every complete subset is sequentially complete and hence every complete topological vector space is sequentially complete. The converse is also true for metrizable topological vector spaces. In fact we have:

PROPOSITION (6): *A metrizable topological vector space is complete if (and only if) it is sequentially complete.*

Proof: Suppose that E is a metrizable topological vector space which is sequentially complete. Let $(x_\lambda)_{\lambda\in\Lambda}$ be a Cauchy net in E

and let (U_n) be a countable base of neighbourhoods of 0. For every $n \in \mathbb{N}$ there exists $\lambda_n \in \Lambda$ such that if $\lambda, \lambda' > \lambda_n$, then $x_\lambda - x_{\lambda'} \in U_n$. Choose $\alpha_n \in \Lambda$ so that $\alpha_n \geq \lambda_i$ for $i = 1,\ldots,n$ and set $y_n = x_{\alpha_n}$. Then (y_n) is a Cauchy sequence in E and hence converges to a point $x \in E$, since E is sequentially complete. We show that the net $(x_\lambda)_{\lambda\in\Lambda}$ converges to x. Let $n \in \mathbb{N}$; there exists $k \in \mathbb{N}$ such that $U_k + U_k \subset U_n$. Since the sequence (y_m) converges to x, we can find an $M \in \mathbb{N}$ such that $y_m - x \in U_k$ whenever $m \geq M$. Let $N = \max(M,k)$ and notice that $\alpha_N \geq \alpha_k \geq \lambda_k$. For $\lambda > \lambda_k$ write:

$$x_\lambda - x = (x_\lambda - y_N) + (y_N - x) = (x_\lambda - x_{\alpha_N}) - (y_N - x).$$

Since $\lambda, \alpha_N > \lambda_k$, $x_\lambda - x_{\alpha_N} \in U_k$ and since $N \geq M$, $y_N - x \in U_k$; therefore $x_\lambda - x \in U_k + U_k \subset U_n$ and the assertion is proved.

0·C.4˙7 Fréchet Spaces

A locally convex space which is metrizable and complete is called a *FRÉCHET SPACE*. Thus a Fréchet space is a locally convex space with a countable base of neighbourhoods of the origin (Subsection 0·C.4˙5) in which every Cauchy sequence is convergent (Subsection 0·C.4˙6). Clearly every Banach space is a Fréchet space.

CHAPTER I

BORNOLOGY

In this Chapter we introduce the basic notions of bornology, bornological vector spaces, bounded linear maps and bornological convergence. We also give many examples, of a general as well as a concrete character, from the usual spaces of Analysis (see also Exercise 1·E.12†). Bounded linear maps are introduced and immediately used for a definition of distributions (Exercise 1·E.12). The remaining Exercises on this Chapter are dedicated to 'von Neumann bornologies', 'bornivorous sets' and bornological convergence for filters.

1:1 DEFINITIONS

1:1·1

A *BORNOLOGY on a set* X is a family $\mathcal{B}$ of subsets of X satisfying the following axioms:

(B.1): *$\mathcal{B}$ is a covering of X, i.e.* $X = \bigcup_{B \in \mathcal{B}} B$;

(B.2): *$\mathcal{B}$ is hereditary under inclusion, i.e. if $A \in \mathcal{B}$ and B is a subset of X contained in A, then $B \in \mathcal{B}$*;

(B.3): *$\mathcal{B}$ is stable under finite union.*

A pair $(X,\mathcal{B})$ consisting of a set X and a bornology $\mathcal{B}$ on X is called a *BORNOLOGICAL SET*, and the elements of $\mathcal{B}$ are called the *BOUNDED SUBSETS* of X.

A *BASE OF A BORNOLOGY* $\mathcal{B}$ on X is any subfamily $\mathcal{B}_0$ of $\mathcal{B}$ such that every element of $\mathcal{B}$ is contained in an element of $\mathcal{B}_0$. A family $\mathcal{B}_0$ of subsets of X is a base for a bornology on X if and only if $\mathcal{B}_0$ covers X and every finite union of elements of $\mathcal{B}_0$ is contained in a member of $\mathcal{B}_0$. Then the collection of those subsets

† I.e. Exercise 1·E.12.

of X which are contained in an element of $\mathcal{B}_0$ defines a bornology $\mathcal{B}$ on X having $\mathcal{B}_0$ as a base. A bornology is said to be a *BORNOLOGY WITH A COUNTABLE BASE* if it possesses a base consisting of a sequence of bounded sets. Such a sequence can always be assumed to be increasing.

1:1·2

Let E be a vector space over $\mathbb{K}$. A bornology $\mathcal{B}$ on E is said to be a *BORNOLOGY COMPATIBLE WITH A VECTOR SPACE STRUCTURE* of E, or to be a *VECTOR BORNOLOGY* on E, if $\mathcal{B}$ is stable under vector addition, homothetic transformations and the formation of circled hulls (*cf.* Section 0·A.3) or, in other words, if the sets $A + B$, λA, $\bigcup_{|\alpha|\leqslant 1} \alpha A$ belong to $\mathcal{B}$ whenever A and B belong to $\mathcal{B}$ and $\lambda \in \mathbb{K}$.

Notice that any hereditary family $\mathcal{B}$ of circled subsets of E satisfying the three conditions above is necessarily stable under finite union: in fact, if $A,B \in \mathcal{B}$, then A and B are circled, hence they contain 0 and, consequently, $A \cup B \subset A + B$.

We call a *BORNOLOGICAL VECTOR SPACE* any pair $(E,\mathcal{B})$ consisting of a vector space E and a vector bornology on E. An equivalent definition (*cf.* Exercise 2·E.1), in complete symmetry with the definitions of topological vector spaces given in Chapter 0, will be given once the notion of 'product bornology' has been introduced (Section 2:2).

1:1·3

A vector bornology on a vector space E is called a *CONVEX VECTOR BORNOLOGY* if it is stable under the formation of convex hulls. Such a bornology is also stable under the formation of disked hulls, since the convex hull of a circled set is circled (Section 0·A.3). A bornological vector space $(E,\mathcal{B})$ whose bornology $\mathcal{B}$ is convex will be called a *CONVEX BORNLOGICAL VECTOR SPACE* or simply a *CONVEX BORNOLOGICAL VECTOR SPACE*.

1:1·4

A *SEPARATED BORNOLOGICAL VECTOR SPACE* $(E,\mathcal{B})$ (or a *SEPARATED BORNOLOGY* $\mathcal{B}$) is one where $\{0\}$ is the only bounded vector subspace of E.

1:2 BOUNDED LINEAR MAPS

1:2·1

Let X and Y be two bornological sets and $u:X \to Y$ a map of X into Y. We say that u is a *BOUNDED MAP* if the image under u of every bounded subset of X is bounded in Y. Obviously the identity map of any bornological set is bounded.

Let X,Y,Z be three bornological sets and $u:X \to Y$, $v:Y \to Z$ be two bounded maps. It is immediate from the definition that the composite map $v\circ u:X \to Z$ is bounded.

A bornology $\mathcal{B}_1$ on a set X is a *FINER BORNOLOGY* than a bornology $\mathcal{B}_2$ on X (or $\mathcal{B}_2$ is a *COARSER BORNOLOGY* than $\mathcal{B}_1$) if the identity map $(X,\mathcal{B}_1) \to (X,\mathcal{B}_2)$ is bounded. This is equivalent to $\mathcal{B}_1 \subset \mathcal{B}_2$.

A *BORNOLOGICAL ISOMORPHISM* between two bornological sets is a bijection u such that both u and its inverse u^{-1} are bounded.

1:2·2

Let, now, E and F be two bornological vector spaces. A *BOUNDED LINEAR MAP* of E into F is any map of E into F which is at the same time linear and bounded. Clearly the composition of two bounded linear maps is a bounded linear map. A trivial example of a bounded linear map is afforded by the identity of any bornological vector space. A bornological isomorphism between two bornological vector spaces is a bornological isomorphism between sets which is also linear.

1:2·3

A *BOUNDED LINEAR FUNCTIONAL* (*FORM*) on a bornological vector space E is a bounded linear map of E into the scalar field $\mathbb{K}$, the latter being endowed with the usual bornology defined by its absolute value (*cf.* Example (1) of Section 1:3 below).

1:3 FUNDAMENTAL EXAMPLES OF BORNOLOGIES

EXAMPLE (1): Let $\mathbb{K}$ be a field with an absolute value (we shall assume that $\mathbb{K}$ is either $\mathbb{R}$ or $\mathbb{C}$). The collection of subsets of $\mathbb{K}$ which are 'bounded' in the usual sense for the absolute value is a convex bornology on $\mathbb{K}$ called the *CANONICAL BORNOLOGY of* $\mathbb{K}$.

EXAMPLE (2): *The Bornology Defined by a Semi-Norm*: Let E be a vector space over $\mathbb{K}$ and let p be a semi-norm on E. A subset A of E is said to be a *SUBSET BOUNDED FOR THE SEMI-NORM* p if $p(A)$ is a bounded subset of $\mathbb{R}$ in the sense of Example (1). The subsets of E which are bounded for the semi-norm p form a convex bornology on E called the *CANONICAL BORNOLOGY OF THE SEMI-NORMED SPACE* (E,p). This bornology is separated if and only if p is a norm. Example (1) is then a particular case of Example (2), which in turn is a particular case of the following general example.

EXAMPLE (3): *The Bornology Defined by a Family of Semi-Norms*: Let E be a vector space and $\Gamma = (p_i)_{i\in I}$ a family of semi-norms on E indexed by a non-empty set I. We agree to say that a subset A of E is a *SUBSET BOUNDED FOR THE FAMILY* Γ *OF SEMI-NORMS* if for every $i \in I$, $p_i(A)$ is bounded in $\mathbb{R}$. The subsets of E which are bounded for the family Γ define a convex bornology on E called the *BORNOLOGY DEFINED BY* Γ. Such a bornology is separated if and only if Γ separates E, i.e. if for every $x \in E$, $x \neq 0$, there exists $i \in I$ such that $p_i(x) \neq 0$. This Example will be generalised in Chapter II to the notion of 'initial bornology'

EXAMPLE (4): *The von Neumann Bornology of a Topological Vector Space*: A *BOUNDED SUBSET OF A TOPOLOGICAL VECTOR SPACE* E is a subset that is absorbed by every neighbourhood of zero. This definition is due to von Neumann (1935). The collection $\mathfrak{B}$ of bounded subsets of E forms a vector bornology on E called the *VON NEUMANN BORNOLOGY* of E or, if no confusion is likely to arise, the *CANONICAL BORNOLOGY* of E. Let us verify that $\mathfrak{B}$ is indeed a vector bornology on E. If $\mathcal{V}$ is a base of circled neighbourhoods of zero in E, it is clear that a subset A of E is bounded if and only if for every $V \in \mathcal{V}$ there exists $\lambda > 0$ such that $A \subset \lambda V$ (Section 0·A.3). Since every neighbourhood of zero is absorbent, $\mathfrak{B}$ is a covering of E. $\mathfrak{B}$ is obviously hereditary and we shall show that it is also stable under vector addition. Let $A, B \in \mathfrak{B}$ and $V \in \mathcal{V}$; there exists $W \in \mathcal{V}$ such that $W + W \subset V$ (Section 0·B.2, Theorem (1)). Since A and B are bounded in E, there exist positive scalars λ and μ such that $A \subset \lambda W$ and $B \subset \mu W$. With $\alpha = \max(\lambda, \mu)$ we have:

$$A + B \subset \lambda W + \mu W \subset \alpha W + \alpha W \subset \alpha(W + W) \subset \alpha V.$$

Finally, since $\mathcal{V}$ is stable under the formation of circled hulls (resp. under homothetic transformations), then so is $\mathfrak{B}$, and we conclude that $\mathfrak{B}$ is a vector bornology on E. If E is locally convex, then clearly $\mathfrak{B}$ is a convex bornology. Moreover, since every topological vector space has a base of closed neighbourhoods of 0, the closure of each bounded subset of E is again bounded. Other properties of the von Neumann bornologies are established in the Exercises.

EXAMPLE (5): *The compact Bornology of a Topological Space*: Let X be a separated topological space. The relatively compact subsets of X form a bornology on X having the family of compact subsets of X as a base. Such a bornology is called the *COMPACT BORNOLOGY of* X. The compact bornology of a separated topological vector space is a vector bornology. In fact, let us denote this bornology by $\mathfrak{B}$. For every scalar $\lambda \in \mathbb{K}$, the map $x \to \lambda x$ of E into E is continuous, hence for every compact set $A \subset E$, λA is compact. Similarly, the continuity of the map $(x,y) \to x + y$ of $E \times E$ into E ensures that the set $A + B$ is compact whenever A and B are compact subsets of E. Finally, for every compact $A \subset E$, the circled hull of A is compact, since it is the image of the set $D \times A$ (D the compact unit ball of $\mathbb{K}$) under the continuous map $(\lambda, x) \to \lambda x$ of $\mathbb{K} \times E$ into E. In general, the compact bornology of of a topological vector space, even a normed one, is not convex (*cf.* Exercise 4·E.9; see, however, Example (10) below). For this reason one often considers the following bornology:

EXAMPLE (6): *The Bornology of Compact Disks of a Topological Vector Space*: A compact disk in a separated topological vector space E is a set which is simultaneously compact and disked. The family $\mathfrak{B}$ of subsets of compact disks of E forms a convex bornology on E. In fact, $\mathfrak{B}$ is a covering of E for, if

$a \in E$, the disked hull of a is the set $\{\lambda a : |\lambda| \leqslant 1\}$ (Proposition (3), Section 0·A.3) and this is a compact disk in E as the image of the unit ball of $\mathbb{K}$ under the continuous map $\lambda \to \lambda a$. Clearly $\mathcal{B}$ is also hereditary and stable under homothetic transformations and the formation of disked hulls. Finally, if A and B are two compact disks, their sum is compact (Example (5)) and a disk; therefore, $\mathcal{B}$ is a convex bornology.

EXAMPLE (7): For two topological vector spaces E and F we *denote by* $L(E,F)$ the *VECTOR SPACE OF ALL CONTINUOUS LINEAR MAPS of* E *into* F. A subset H of $L(E,F)$ is called an *EQUICONTINUOUS SUBSET* if for every neighbourhood V of zero in F the set $H^{-1}(V) = \bigcap_{u \in H} u^{-1}(V)$ is a neighbourhood of zero in E. In this definition it is clearly enough to assume that V belongs to a base of neighbourhoods of zero in F.

The family $\mathcal{K}$ *of equicontinuous subsets of* $L(E,F)$ *is a vector bornology on* $L(E,F)$. *This bornology is called the EQUICONTINUOUS BORNOLOGY of* $L(E,F)$ *and is a convex bornology if* F *is locally convex.* We shall now prove this assertion.

Since every element of $L(E,F)$ is continuous by definition, $\mathcal{K}$ covers $L(E,F)$ and is also clearly hereditary. Let us show that $\mathcal{K}$ is stable under vector addition. Let $\mathcal{V}$ be a base of circled neighbourhoods of zero in F and let $H_1, H_2 \in \mathcal{K}$. If $V \in \mathcal{V}$, there exists $W \in \mathcal{V}$ such that $W + W \subset V$. By virtue of the equicontinuity of H_1 and H_2, the sets ${H_1}^{-1}(W)$ and ${H_2}^{-1}(W)$ are neighbourhoods of zero in E. Now $(H_1 + H_2)^{-1}(V)$ contains ${H_1}^{-1}(W) \cap {H_2}^{-1}(W)$ and hence is a neighbourhood of zero in E. Thus $H_1 + H_2 \in \mathcal{K}$ since V was arbitrary in $\mathcal{V}$. The family $\mathcal{K}$ is also stable under homothetic transformations since for every $H \in \mathcal{K}$, $\lambda \in \mathcal{K}$ and $V \in \mathcal{V}$ we have:

$$(\lambda H)^{-1}(V) = \frac{1}{\lambda} H^{-1}(V) = \frac{1}{\lambda}\left(\bigcap_{u \in H} u^{-1}(V)\right),$$

which shows the set $(\lambda H)^{-1}$ to be a neighbourhood of zero in E. Finally, $\mathcal{K}$ is stable under the formation of circled hulls. In fact, if $H \in \mathcal{K}$ and if H_1 is the circled hull of H in $L(E,F)$, then for every $V \in \mathcal{V}$ we have ${H_1}^{-1}(V) \supset H^{-1}(V)$. Thus $\mathcal{K}$ is a vector bornology on $L(E,F)$.

Suppose, now, F to be locally convex. If V is a disked neighbourhood of zero in F and $H \in \mathcal{K}$, then $(\Gamma(H))^{-1}(V) \supset H^{-1}(V)$: in fact, if $x \in E$ is such that $u(x) \in V$ for every $u \in H$, then, since V is a disk, $\sum \lambda_i h_i(x) \in V$ for every finite family (λ_i) of scalars such that $\sum |\lambda_i| \leqslant 1$ and for every finite family (h_i) of arbitrary elements of H. Thus $v(x) \in V$ whenever $v \in \Gamma(H)$, which shows that $\mathcal{K}$ is a convex bornology.

EXAMPLE (8): *The Natural Bornology*: Let X be a set, σ a family of subsets of X and $(F,\mathcal{B})$ a bornological set. A family B of maps of X into F is called σ-*BOUNDED* if $B(A) = \bigcup_{u \in B} u(A)$ is bounded in $(F,\mathcal{B})$ for every $A \in \sigma$. Let H be a subset of the set F^X of all maps of X into F. If every point of H is σ-bounded, the σ-bounded subsets of H define a bornology on H called the *σ-BORNOLOGY*. When the pair (X,σ) is a bornological set, the σ-bornology on a subset H of F^X is called the *NATURAL BORNOLOGY* on H. A subset of H which is bounded for the natural bornology will then be said to be *EQUIBOUNDED* on every bounded subset of (X,σ).

EXAMPLE (9): *The Precompact Bornology of a Topological Vector Space:*

(a): A subset A of a topological vector space E is called *PRECOMPACT* if for every neighbourhood V of zero in E, there exist finitely many points $a_1,a_2,\ldots,a_n$ of E such that $A \subset \bigcup_{i=1}^{n} (a_i + V)$. It is clear that every compact subset of E is precompact, that the union of two precompact sets is precompact and that so is every subset of a precompact set. *Hence the family $\mathcal{P}$ of all precompact subsets of E is a bornology on E. Moreover, $\mathcal{P}$ is a vector bornology.* In fact, let $A,B \in \mathcal{P}$ and let V,W be neighbourhoods of zero in E such that $W + W \subset V$. Then:

$$A \subset \bigcup_{i=1}^{n} (a_i + W) \quad \text{and} \quad B \subset \bigcup_{j=1}^{n} (b_j + W),$$

with $a_i,b_j \in E$ and $n,m \in \mathbb{N}$; hence:

$$A + B \subset \bigcup_{i,j} (a_i + b_j + W + W) \subset \bigcup_{i,j} (a_i + b_j + V),$$

and $A + B$ is precompact. Since λA is precompact for every precompact set A and scalar λ, it remains to show that the circled hull of a precompact set is again precompact. Now if A is precompact and V is a circled neighbourhood of zero in E, the circled hull of A is contained in $M + V$, where M is the circled hull of a finite set. Hence it suffices to show that the circled hull of a finite set N is precompact. Since every finite union of precompact sets is precompact, we may assume that N consists of a single point $a \in E$. The circled hull of N is the set $Da = \{\lambda a;\ \lambda \in \mathbb{K},\ |\lambda| \leqslant 1\}$, where $D = \{\lambda \in \mathbb{K};\ |\lambda| \leqslant 1\}$, whence is the image of D under the continuous linear map $\lambda \to \lambda a$ of $\mathbb{K}$ into E. Since D, being compact in $\mathbb{K}$, is precompact, the precompactness of Da is a consequence of the following general result:

(b): *Let E,F be two topological vector spaces and let $u:E \to F$ be a continuous linear map. If A is a precompact set in E, then $u(A)$ is a precompact set in F.*

In fact, let V be a neighbourhood of zero in F. Since u is continuous, $W = u^{-1}(V)$ is a neighbourhood of zero in E and hence $A \subset A_0 + W$, A_0 being a finite subset of E. Consequently:

$$u(A) \subset u(A_0) + u(W) \subset u(A_0) + V,$$

with $u(A_0)$ a finite set in F. Thus $u(A)$ is precompact.

We now give some further properties of precompact sets.

(c): *In a topological vector space E the closure of a precompact set is precompact.* This follows from the fact that E has a base of closed neighbourhoods of zero.

(d): *In a separated locally convex space the precompact bornology is convex.* We shall show directly that the disked hull of a precompact set is precompact. We begin by showing that the disked hull of a finite set is precompact. Let $\{a_1,\ldots,a_n\}$ be a finite subset of E, let C be its disked hull and let $B = \{(\lambda_1,\ldots,\lambda_n)\in\mathbb{K}^n;\ \sum_{i=1}^{n} |\lambda_i| \leqslant 1\}$. Since C is the image of B under the continuous map $(\lambda_1,\ldots,\lambda_n) \to \sum_{i=1}^{n} \lambda_i a_i$ and B is compact in $\mathbb{K}^n$, C is compact in E. Let, now, A be a precompact subset of E and let V be a disked neighbourhood of zero in E. Then $A \subset \bigcup_{i=1}^{n} (a_i + \frac{1}{2}V)$ with $\{a_1,\ldots,a_n\} \subset E$. Since the disked hull M of $\{a_1,\ldots,a_n\}$ is compact, whence precompact in E, there exists a finite set $\{b_1,\ldots,b_m\} \subset E$ such that $M \subset \bigcup_{i=1}^{m} (b_i + \frac{1}{2}V)$. Now the disked hull A_1 of A is contained in $M + \frac{1}{2}V$, for $\frac{1}{2}V$ is disked. Thus:

$$A_1 \subset \bigcup_{i=1}^{m} (b_i + \tfrac{1}{2}V) + \tfrac{1}{2}V = \bigcup_{i=1}^{m} (b_i + V),$$

and the assertion is proved.

(e): *In every topological vector space the precompact bornology is finer than the von Neumann bornology.* This is an immediate consequence of the definitions.

EXAMPLE (10): *The Compact Bornology of a Banach Space*: We have stated earlier that the compact bornology of a locally convex space, even normed, is not convex in general (for a counter-example, see Exercise 4·E.9). However, the compact bornology is convex in every separated, complete, locally convex space (and, more generally, in every separated locally convex space whose bounded closed sets are complete). We show this in the case of a Banach space E. It suffices to prove that the closed disked hull B of a compact subset A of E is compact. Appealing to the characterisation of compact sets in a metric space (J. Dieudonné [*1*], §16, Proposition 3.16.1), we have to show that B is precompact and complete. Note that the definition of precompactness given in J. Dieudonné [*1*], (§16) and that given in Example (9) above coincide in the case of normed spaces. Now B is closed in E and hence complete. Moreover, B, being the closed disked hull of a precompact set, is precompact by (a,c,d) of Example (9). Thus B is compact.

Therefore, *the compact bornology of a Banach space E is a convex bornology*. If E is infinite-dimensional, there is no vector topology on E whose von Neumann bornology coincides with the compact bornology of E (*cf.* Exercises 1·E.4,13).

1:4 BORNOLOGICAL CONVERGENCE

In every bornological vector space a notion of convergence can be introduced which depends only upon the bornology of the space. For convex bornologies, this convergence reduces to convergence in a normed space, and this fact is of considerable interest in many situations.

1:4·1

DEFINITION (1): *Let E be a bornological vector space. A sequence (x_n) in E is said to* CONVERGE BORNOLOGICALLY *to* 0 *if there exist a circled bounded subset B of E and a sequence (λ_n) of scalars tending to* 0*, such that $x_n \in \lambda_n B$ for every integer $n \in \mathbb{N}$.*

For historical reasons, bornological convergence is also called *MACKEY-CONVERGENCE* after G.W. Mackey, who was the first to study systematically, since 1942, such a notion of convergence in the particular context of its theory of 'linear systems'.

One usually writes $x_n \xrightarrow{M} 0$ to express the fact that the sequence (x_n) converges bornologically to 0. We then say that a *sequence (x_n) converges bornologically to a point $x \in E$* if $(x_n - x) \xrightarrow{M} 0$, and we write $x_n \xrightarrow{M} x$.

Clearly the relations $x_n \xrightarrow{M} x$, $y_n \xrightarrow{M} y$ in E and $\lambda_n \to \lambda$ in $\mathbb{K}$ imply that $(x_n + y_n) \xrightarrow{M} (x + y)$ and $\lambda_n x_n \xrightarrow{M} \lambda x$. It is also evident that every bornologically convergent sequence is bounded and that the image of a bornologically convergent sequence under a bounded linear map is again a bornologically convergent sequence.

We shall now give several characterisations of bornological convergence.

1:4·2

PROPOSITION (1): *Let E be a bornological vector space and let (x_n) be a sequence in E. The following assertions are equivalent:*

(i): *The sequence (x_n) converges bornologically to* 0*;*

(ii): *There exists a circled bounded set $B \subset E$ and a decreasing sequence (α_n) of positive real numbers, tending to* 0*, such that $x_n \in \alpha_n B$ for every $n \in \mathbb{N}$;*

(iii): *There exists a circled bounded set $B \subset E$ such that, given any $\varepsilon > 0$, we can find an integer $N(\varepsilon)$ for which $x_n \in \varepsilon B$ whenever $n \geqslant N(\varepsilon)$.*

If the bornology of E is convex, then (i,ii,iii) *are also equivalent to the following:*

(iv): *There exists a bounded disk $B \subset E$ such that (x_n) is contained in the semi-normed space E_B and converges to 0 in E_B.*

Proof: (i) => (ii): For any integer $p \in \mathbb{N}$ there exists $N_p \in \mathbb{N}$ such that if $n \geqslant N_p$, then $\lambda_n \leqslant 1/p$; hence $\lambda_n B \subset (1/p)B$, since B is circled. We may assume that the sequence N_p is strictly increasing, and, for $N_p \leqslant k < N_{p+1}$, we let $\alpha_k = 1/p$. Then the sequence (α_k) satisfies the conditions of assertion (ii). Clearly (ii) => (iii). To show that (iii) => (i), we let, for every $n \in \mathbb{N}$:

$$\varepsilon_n = \inf\{\varepsilon > 0;\ x_n \in \varepsilon B\} \quad \text{and} \quad \lambda_n = \varepsilon_n + \frac{1}{n}.$$

Then the sequence (λ_n) converges to 0 and $x_n \in \lambda_n B$ for every $n \in \mathbb{N}$. Thus the assertions (i,ii,iii) are equivalent.

Suppose now that the bornology of E is convex. Clearly (iv) implies (i) with $\lambda_n = p_B(x_n)$ and p_B the gauge of B, whilst (ii) implies that $x_n \in E_B$ and $p_B(x_n) \leqslant \alpha_n \to 0$. The proof of the Proposition is now complete.

1:4·3 Relationship between Bornological and Topological Convergence

Let E be a topological vector space. With a little abuse of language, we say that a sequence (x_n) of points of E converges bornologically to x *in* E if it converges bornologically to x when E is endowed with its von Neumann bornology.

Since every bounded subset of E is absorbed by each neighbourhood of 0, every bornologically convergent sequence is topologically convergent. The converse is false in general (Exercise 1·E.14), but is true in 'all' the 'usual' spaces encountered in Analysis, as shown by the following two Propositions:

PROPOSITION (2): *Let E be a separated topological vector space satisfying the following condition:*

(S): *For every compact set $K \subset E$ there exists a bounded disk $B \subset E$ such that K is compact in E_B.*

Then every topologically convergent sequence in E is also bornologically convergent.

Proof: If (x_n) converges topologically to 0 in E, then the set $A = (x_n) \cup \{0\}$ is compact in E and since, by Condition (S), compact in a suitable space E_B. Since the canonical embedding $E_B \to E$ is continuous, the topologies of E and E_B coincide on A and, therefore, (x_n) converges to 0 in E_B.

In particular, let E be a separated locally convex space. If every bounded subset of E is relatively compact in a space E_B, with B a bounded disk in E, then every topologically convergent sequence in E is bornologically convergent.

We shall see in Chapter VIII that this condition simply expresses the fact that the von Neumann bornology of E is of a particular type called '*Schwartz bornology*', the reason for this name being that the above condition is satisfied by the principal spaces of L. Schwartz's theory of distributions.

PROPOSITION (3): *In a metrizable topological vector space (locally convex or not) every topologically convergent sequence is bornologically convergent.*

Proof: Let (V_n) be a countable base of neighbourhoods of 0 in E such that $V_n \supset V_{n+1}$ for every $n \in \mathbb{N}$ and let $A = (x_k)$ be a sequence in E which converges to 0 topologically. We are going to prove the following assertion:

(*): '*There exists a circled, bounded set B such that, for every $\varepsilon > 0$, there is an integer $m \in \mathbb{N}$ for which $A \cap V_m \subset \varepsilon B$*'.

By Proposition (1)(iii) of Subsection 1:4·2 this assertion implies that the sequence (x_k) converges bornologically to 0: in fact, if $N(\varepsilon)$ is a positive integer such that $x_k \in V_m$ whenever $k \geqslant N(\varepsilon)$, then $x_k \in A \cap V_m \subset \varepsilon B$ for $k \geqslant N(\varepsilon)$. Thus, it suffices to prove assertion (*).

Since the sequence A converges topologically to zero, it is absorbed by every neighbourhood of zero. Hence, for every $n \in \mathbb{N}$, there exists a positive real number λ_n such that $A \subset \lambda_n V_n$. It follows that:

$$A \subset \bigcap_{n=1}^{\infty} \lambda_n V_n .$$

Let (α_n) be a sequence of strictly positive real numbers converging to 0, let $\mu_n = \lambda_n/\alpha_n$ and consider the set:

$$B = \bigcap_{n=1}^{\infty} \lambda_n V_n .$$

Clearly B is bounded in E and we claim that B is the set whose existence is asserted by (*). Let $\varepsilon > 0$ be given. Since the sequence $\mu_n/\lambda_n = 1/\alpha_n$ tends to $+\infty$, there is an integer $\ell \in \mathbb{N}$ such that, for $n \geqslant \ell$, $\mu_n/\lambda_n \geqslant 1/\varepsilon$, i.e. $\lambda_n \leqslant \varepsilon\mu_n$. Then, since $A \subset \bigcap_{n=1}^{\infty} \lambda_n V_n$, $A \subset \varepsilon\mu_n V_n$ for $n \geqslant \ell$. But the set $\bigcap_{n<\ell} \varepsilon\mu_n V_n$ is a neighbourhood of 0 and hence there exists an integer $m \in \mathbb{N}$ such that $V_m \subset \bigcap_{n<\ell} \varepsilon\mu_n V_n$. Thus $A \cap V_m \subset \varepsilon\mu_n V_n$ for every $n \in \mathbb{N}$ and, therefore,

$$A \cap V_m \subset \varepsilon \bigcap_{n=1}^{\infty} \mu_n V_n = \varepsilon B.$$

REMARK: For locally convex spaces, another proof of the above Proposition can be found in the Exercises.

1:4·4 Uniqueness of Bornological Limits

PROPOSITION (4): *A bornological vector space E is separated if and only if every bornologically convergent sequence in E has a unique limit.*

Proof: Necessity: Let E be a separated bornological vector space. If a sequence (x_n) in E converges bornologically to x and y, then the sequence $x_n - x_n = 0$ converges to $x - y$. Thus it suffices to show that the limit z of the sequence $(z_n = 0)$ must be the element 0. Let (λ_n) be a sequence of real numbers tending to 0 and let B be a bounded subset of E such that $z - z_n = z \in \lambda_n B$ for every integer $n \geqslant 1$. If $z \neq 0$, then the line spanned by z (i.e. the subspace $\mathbb{K}z$) is contained in B, contradicting the hypothesis that E is separated.

Sufficiency: Assume uniqueness of limts and suppose that there exists an element $z \neq 0$ such that the line spanned by z is bounded. Then we can find a bounded set $B \subset E$ such that $z \in (1/n)B$ for every $n \geqslant 1$ and hence the sequence $(z_n = z)$ converges to 0. But clearly this sequence also converges to z, whence, by uniqueness of limits $z = 0$ and we have reached a contradiction.

CHAPTER II

FUNDAMENTAL BORNOLOGICAL CONSTRUCTIONS

This Chapter gives the fundamental methods for constructing new bornologies from given ones. These methods are standard in Functional Analysis and consist in forming products, subspaces, projective limits, quotients, inductive limits and direct sums. In the case of vector bornologies, conditions are given ensuring that the new bornologies thus obtained are separated. In Section 2:13 convex bornological spaces are characterised as bornological limits of semi-normed spaces. This enables us to make clear the *essential* difference between the 'bornological structure' and the 'topological structure' of a vector space: the former is a collection of 'internal pieces' each of which is a normed space, whilst the latter is a collection of 'external hulls' each of which is a normed space. Hence the two fundamental methods of investigation: by 'union of normed spaces' and by 'intersection of normed spaces'.

In the Exercises the bornologies constructed by the methods indicated above are compared with the von Neumann bornologies associated with the locally convex topologies constructed by analogous methods (example: quotient bornology and von Neumann bornology of a quotient topology). An interpretation of the bornologies of the spaces of differentiable functions in terms of initial or final bornologies is also given in the Exercises.

2:1 INITIAL BORNOLOGIES

THEOREM (1): *Let I be a non-empty set, let $(X_i, \mathcal{B}_i)_{i\in I}$ be a family of bornological sets indexed by I and let X be a set. Suppose that, for every $i \in I$, a map $u_i : X \to X_i$ is given. Consider the set $\mathcal{B}$ of all subsets A of X having the following property:*

'For every $i \in I$, $u_i(A)$ is bounded in X_i'.

Then:

(i): *$\mathfrak{B}$ is a bornology on X and is the coarsest bornology on X for which each map u_i is bounded;*

(ii): *If X is a vector space and if for every $i \in I$, X_i is a vector space, $\mathfrak{B}_i$ is a vector (resp. convex) bornology on X_i and the map u_i is linear, then $\mathfrak{B}$ is a vector (resp. convex) bornology on X.*

DEFINITION (1): *The bornology $\mathfrak{B}$ on X defined by Theorem* (1) *is called the* INITIAL BORNOLOGY *on X for the maps u_i.*

REMARK (1): Let Y be a bornological set and let X be endowed with the initial bornology for the maps u_i. Then a map $u: Y \to X$ is bounded if and only if $u_i \circ u$ is bounded for every $i \in I$.

2:1·1 Base of an Initial Bornology

With the notation of Theorem (1), let:

$$u_i^{-1}(\mathfrak{B}_i) = \{u_i^{-1}(A): A \in \mathfrak{B}_i\} \qquad \text{for every } i \in I.$$

Then the family $\mathfrak{B}_0 = \bigcap_{i \in I} u_i^{-1}(\mathfrak{B}_i)$ is a base of the initial bornology $\mathfrak{B}$ on X for the maps u_i. In fact, on the one hand, every element A of $\mathfrak{B}_0$ is evidently bounded for $\mathfrak{B}$, since $u_i(A) \in \mathfrak{B}_i$ for each $i \in I$. On the other hand, if $A \in \mathfrak{B}$, then, for every $i \in I$, $u_i(A)$ is bounded in X_i and hence there exists $B_i \in \mathfrak{B}_i$ such that $u_i(A) \subset B_i$, i.e. $A \subset u_i^{-1}(B_i)$. Thus the intersection of the sets $u_i^{-1}(B_i)$ belongs to $\mathfrak{B}_0$ and contains A, and the assertion follows.

The most important particular cases of initial bornologies are given in the following Sections 2:2-5.

2:2 PRODUCT BORNOLOGIES

DEFINITION (1): *Let $(X_i, \mathfrak{B}_i)_{i \in I}$ be a family of bornological sets indexed by a non-empty set I and let $X = \prod_{i \in I} X_i$ be the product of the sets X_i. For every $i \in I$, let $p_i: X \to X_i$ be the canonical projection. Then the* PRODUCT BORNOLOGY *on X is the initial bornology on X for the maps p_i.*

The set X, endowed with the product bornology, is called the *BORNOLOGICAL PRODUCT* of the sets $(X_i, \mathfrak{B}_i)$.

PROPOSITION (1): *With the notation of Definition* (1), *the product bornology on X has a base consisting of sets of the form $B = \prod_{i \in I} B_i$, where $B_i \in \mathfrak{B}_i$ for all $i \in I$.*

Proof: A set of the form $B = \prod_{i \in I} B_i$ is clearly bounded for the product bornology, since its projections are bounded. Conversely if A is a subset of X which is bounded for the product bornology, let $A_i = p_i(A)$ for every $i \in I$, and $B = \prod_{i \in I} A_i$. Then each set A_i is bounded in X_i and $A \subset B$.

REMARK (1): If the X_i's are vector spaces and X is their product regarded as a vector space, then the projections $p_i : X \to X_i$ are linear. Thus by Theorem (1) of Section 2:1 the product bornology is a vector (resp. convex) bornology if all the bornologies $\mathcal{B}_i$ are also.

2:3 INDUCED BORNOLOGIES: BORNOLOGICAL SUBSPACES

DEFINITION (1): *Let $(X,\mathcal{B})$ be a bornological set, let Y be a subset of X and let $u : Y \to X$ be the canonical embedding. Then the bornology induced on Y by $(X,\mathcal{B})$ is the initial bornology on Y for the map u.*

The set Y, endowed with the bornology induced by $(X,\mathcal{B})$, is said to be a *BORNOLOGICAL SUBSET* of $(X,\mathcal{B})$. If $(X,\mathcal{B})$ is a bornological vector space, the induced bornology on Y is necessarily a vector bornology, and Y is then called a *BORNOLOGICAL SUBSPACE* of X.

PROPOSITION (1): *With the notation of Definition* (1), *a base for the bornology induced on Y by $(X,\mathcal{B})$ is given by the family $\{A \cap Y : A \in \mathcal{B}\}$.*

This Proposition is an immediate consequence of the definitions.

2:4 BORNOLOGIES GENERATED BY A FAMILY OF SUBSETS

DEFINITION (1): *Let X be a set and let $(\mathcal{B}_i)_{i \in I}$ be a family of bornologies on X indexed by a non-empty set I. For every $i \in I$, let u_i be the identity map of X onto $(X,\mathcal{B}_i)$. The INTERSECTION OF THE BORNOLOGIES $\mathcal{B}_i$ is the initial bornology on X for the maps u_i.*

Evidently $\bigcap_{i \in I} \mathcal{B}_i$ is a base for such a bornology. If X is a vector space and if, for every $i \in I$, $\mathcal{B}_i$ is a vector (resp. convex) bornology, then the intersection bornology is also a vector (resp. convex) bornology by Theorem (1) of Section 2:1.

DEFINITION (2): *Let X be a set (resp. vector space) and let $\mathcal{A}$ be a family of subsets of X. The BORNOLOGY (resp. VECTOR BORNOLOGY, resp. CONVEX BORNOLOGY) GENERATED BY $\mathcal{A}$ is the intersection of all bornologies (resp. vector bornologies, resp. convex bornologies) containing $\mathcal{A}$.*

Note that there always exists a bornology which contains $\mathcal{A}$, namely the bornology $\mathcal{B} = \mathcal{P}(X)$ whose bounded sets are all the subsets of X and, if X is a vector space, this bornology is convex.

2:5 BORNOLOGICAL PROJECTIVE LIMITS

2:5·1 Bornological Projective Systems

Let I be a non-empty directed set and let (X_i, u_{ij}) be a projective system of sets, indexed by I (*cf.* Section 0·A.2), such that for every $i \in I$, X_i is a bornological set with bornology $\mathcal{B}_i$. The system (X_i, u_{ij}) is called a *PROJECTIVE SYSTEM OF BORNOLOGICAL SETS* if the maps $u_{ij}: X_j \to X_i$ are bounded whenever $i \leqslant j$. If the X_i's are bornological vector spaces (resp. convex bornological spaces) and all the maps u_{ij} are bounded and linear, then the system (X_i, u_{ij}) is called a *PROJECTIVE SYSTEM OF BORNOLOGICAL VECTOR SPACES* (resp. *OF CONVEX BORNOLOGICAL SPACES*).

2:5·2 Projective Limit Bornologies

Let (X_i, u_{ij}) be a projective system of bornological sets (resp. bornological vector spaces, resp. convex bornological spaces) and let X be the set (resp. the vector space) which is the projective limit of the system (X_i, u_{ij}) (Section 0·A.2). For every $i \in I$, denote by $\mathcal{B}_i$ the bornology of X_i and by u_i the canonical projection of X onto X_i. The *PROJECTIVE LIMIT BORNOLOGY* on X with respect to the bornologies $\mathcal{B}_i$ is the initial bornology on X for the maps u_i and X, endowed with such a bornology, is called the *BORNOLOGICAL PROJECTIVE LIMIT* of the bornological projective system (X_i, u_{ij}). We shall then write:

$$X = \varprojlim_{i \in I} (X_i, u_{ij}).$$

REMARK (1): The reader can verify in the Exercises that the product bornology is a particular case of a projective limit bornology.

2:6 FINAL BORNOLOGIES

THEOREM (1): *Let I be a non-empty set, let $(X_i, \mathcal{B}_i)_{i \in I}$ be a family of bornological sets and let X be a set. Suppose that, for every $i \in I$, a map $v_i: X_i \to X$ is given and let $\mathcal{B}$ be the bornology on X generated by the family $\mathcal{A} = \bigcup_{i \in I} v_i(\mathcal{B}_i)$. Then:*

(i): *$\mathcal{B}$ is the finest bornology on X for which each map v_i is bounded;*

(ii): *If X is a vector space and if, for every $i \in I$, X_i is a vector space, $\mathcal{B}_i$ is a vector (resp. convex) bornology and the map v_i is linear, then the vector (resp. convex) bornology on X generated by $\mathcal{A}$ is the finest vector (resp. convex) bornology on X for which all the maps v_i are bounded.*

This Theorem is a straightforward consequence of the definitio

DEFINITION (1): *The bornology $\mathcal{B}$ on X constructed in Theorem (1)(i) (resp. Theorem (1)(ii)) is called the FINAL BORNOLOGY (resp. FINAL VECTOR BORNOLOGY, resp. FINAL CONVEX BORNOLOGY) on X for the maps v_i.*

REMARK (1): Let Y be a bornological set and let X be endowed with the final bornology for the maps v_i. Then a map $v:X \to Y$ is bounded if and only if $v \circ v_i$ is bounded for every $i \in I$.

We now proceed to give, in the following Sections 2:7,8,9, the three most important cases of final bornologies.

2:7 QUOTIENT BORNOLOGIES

DEFINITION (1): *Let $(X,\mathcal{B})$ be a bornological set and let $\varphi:X \to Y$ be a map of X onto a set Y. Then the image bornology of $\mathcal{B}$ under φ is the final bornology on Y for the map φ.*

Since φ is onto, it is clear, by virtue of Theorem (1) of Section 2:6, that $\varphi(\mathcal{B})$ is a base for the image bornology of $\mathcal{B}$ under φ. Suppose that $(X,\mathcal{B})$ is a bornological vector space (resp. convex bornological space), that Y is a vector space and that φ is linear; then it follows from the definitions that the image bornology is a vector (resp. convex) bornology.

Let now Y be the quotient of the set X by an arbitrary equivalence relation, φ denoting the canonical map of X onto Y. Then Y, equipped with the image bornology of $\mathcal{B}$ under φ, is called the *BORNOLOGICAL QUOTIENT* of $(X,\mathcal{B})$ and the bornology $\varphi(\mathcal{B})$ is called the *QUOTIENT BORNOLOGY* of $\mathcal{B}$ by the given equivalence relation.

If we take for X a bornological vector space (resp. convex bornological space) E and for Y the quotient E/F, where F is a vector subspace of E, then the image bornology of $\mathcal{B}$ is a vector (resp. convex) bornology, since in this case the canonical map is linear.

2:8 BORNOLOGICAL INDUCTIVE LIMITS

2:8·1 Bornological Inductive Limits

Let I be a non-empty directed set and let (X_i,v_{ji}) be an inductive system of sets, indexed by I (*cf.* Section 0·A.1), such that for every $i \in I$, X_i is a bornological set with bornology $\mathcal{B}_i$. The system (X_i,v_{ji}) is called an *INDUCTIVE SYSTEM OF BORNOLOGICAL SETS* if the maps $v_{ji}:X_i \to X_j$ are bounded whenever $i \leqslant j$. If the X_i's are bornological vector spaces (resp. convex bornological spaces) and all the maps v_{ji} are bounded and linear, then the system (X_i,v_{ji}) is called an *INDUCTIVE SYSTEM OF BORNOLOGICAL VECTOR SPACES* (resp. *OF CONVEX BORNOLOGICAL SPACES*).

2:8·2 Inductive Limit Bornologies

Let (X_i,v_{ji}) be an inductive system of bornological sets (resp. bornological vector spaces, resp. convex bornological spaces) and let X be the set (resp. the vector space) which is the inductive limit of the system (X_i, v_{ji}) (Section 0·A.1). For every $i \in I$, denote by $\mathcal{B}_i$ the bornology of X_i and by v_i the canonical map of X_i into X. The *INDUCTIVE LIMIT BORNOLOGY* on X with respect to bornologies $\mathcal{B}_i$ is the final bornology on X for the maps v_i. For every $i \in I$, let $v_i(\mathcal{B}_i) = \{v_i(A): A \in \mathcal{B}_i\}$. Then the family $\mathcal{B} = \bigcup_{i \in I} v_i(\mathcal{B}_i)$ is precisely the final bornology on X. In fact, we already know from Theorem (1) of Section 2:6 that $\mathcal{B}$ generates the finaly bornology; however, $X = \bigcup_{i \in I} v_i(\mathcal{B}_i)$ and hence $\mathcal{B}$ is indeed a bornology. It follows that, if (X_i, v_{ji}) is an inductive system of bornological vector spaces (resp. of convex bornological spaces), then the inductive limit bornology on X is *necessarily a vector* (resp. *convex*) *bornology*.

When given the inductive limit bornology, X is called the *BORNOLOGICAL INDUCTIVE LIMIT* of the bornological inductive system (X_i, v_{ji}) and *denoted by*:

$$X = \varinjlim_{i \in I} (X_i, v_{ji}).$$

2:9 BORNOLOGICAL DIRECT SUMS: FINITE-DIMENSIONAL BORNOLOGIES

2:9·1

DEFINITION (1): *Let I be a non-empty set of indices, let $(E_i)_{i \in I}$ be a family of bornological vector spaces and let E be the vector space direct sum of the E_i's. For every $i \in I$, denote by $\mathcal{B}_i$ the bornology of E_i and by v_i the canonical map of E_i into E. The direct sum bornology with respect to the bornologies $\mathcal{B}_i$ is the final vector bornology on E for the maps v_i.*

Equipped with the direct sum bornology, E is called the *BORNOLOGICAL DIRECT SUM* of the spaces E_i and we write:

$$E = \bigoplus_{i \in I} E_i.$$

PROPOSITION (1): *With the notation of Definition* (1), *the family of subsets of E of the form:*

$$B = \bigoplus_{i \in I} B_i,$$

(finite sum of bounded sets $B_i \subset E_i$), is a base for the direct sum bornology on E.

Proof: The family of all such finite sums covers E, is stable under the formation of finite sums and scalar multiples, and consists of circled sets, whence is a base of a vector bornology $\mathcal{B}$

for which the maps v_i are continuous. It follows that $\mathcal{B}$ contains the vector bornology generated by $\bigcup_{i\in I} v_i(\mathcal{B}_i)$. However, the latter contains all finite sums of the form $\bigoplus_{i\in I} B_i$, since the maps v_i are injective and, therefore, the two bornologies coincide.

2:9˙2 Direct Sum Bornology and Product Bornology

Let $(E_i)_{i\in I}$ be a family of bornological vector spaces, let $F = \prod_{i\in I} E_i$ be their bornological product and let $E = \bigoplus_{i\in I} E_i$ be their bornological direct sum. It is clear that E is a vector subspace of F and *on E the direct sum bornology is always finer than the bornology induced by F* (*cf.* Exercise 2·E.2). *These two bornologies coincide, however, if I is finite.* In fact, in this case, and we may assume that $(E_i)_{i\in I}$ consists of only two elements E_1 and E_2, it is immediately seen that the canonical map $(x_1,x_2) \to x_1 + x_2$ of $E_1 \times E_2$ onto $E_1 \oplus E_2$ is a bornological isomorphism.

2:9˙3 Direct Sums as Special Inductive Limits

With the notation of Subsection 2:9˙2 let $\mathfrak{F}(I)$ be the set of all finite subsets of I ordered by inclusion and, for every $J \in \mathfrak{F}(I)$, let $E_J = \bigoplus_{i\in J} E_i$ be the bornological direct sum of the spaces $(E_i)_{i\in J}$. For $J \subset J'$ denote by $u_{J'J}$ the canonical embedding of E_J in $E_{J'}$. Then $(E_J, u_{J'J})$ is an inductive system of bornological vector spaces and $E = \varinjlim(E_J, u_{J'J})$.

2:9˙4 The Finite-Dimensional Bornology

Let E be a vector space over $\mathbb{K}$. Algebraically, E is isomorphic to the direct sum $\mathbb{K}^{(I)}$ of copies of $\mathbb{K}$ indexed by some set I. Hence, if we identify E and $\mathbb{K}^{(I)}$, we can consider on E the direct sum bornology with respect to copies of the canonical bornology of $\mathbb{K}$ and the space E then becomes the bornological inductive limit of its finite-dimensional subspaces $\mathbb{K}^n$, $n \in \mathbb{N}$. This bornology is called the *FINITE-DIMENSIONAL BORNOLOGY* on E. It is the finest vector bornology on E and is always convex.

2:9˙5 Bornologically Complementary Subspaces

Let E be a bornological vector space and let M,N be subspaces of E such that E is the algebraic direct sum of M and N. We give M and N the bornology induced by E and we say that M and N are *BORNOLOGICALLY COMPLEMENTARY SUBSPACES* in E and that M (resp. N) is a *BORNOLOGICAL COMPLEMENT* of N (resp. M) in E if E is the bornological direct sum of M and N. For this, *a sufficient (and also necessary) condition is that one of the projections on the subspaces M and N be bounded. Then both projections will be bounded.* In fact, let E be the bornological direct sum of M and N, and let p_M (resp. p_N) be the projection of E onto M (resp. N). Since a base for the bornology of E consists of sets of the form $A \oplus B$, with A and B bounded in M and N respectively, the projections p_M and p_N are bounded. Conversely, if, for example, p_N is bounded, then p_N is also bounded, since $p_M + p_N$ is the identity of E. Now if C is a bounded set in E, then $p_M(C)$ and $p_N(C)$ are bounded in M and N respectively and, since $C \subset p_M(C) \oplus p_N(C)$, C is bounded for the direct sum bornology. This shows that the bornology of E is finer than the direct sum bornology. But the latter is evidently finer than the bornology of E, for the sum of two bounded subsets of E is bounded. Therefore, the two bornologies are the same.

2:10 STABILITY OF THE SEPARATION PROPERTY

In this Section we investigate how the property of a bornology being separated behaves with respect to the fundamental constructions described above. As usual, I is a non-empty set of indices.

PROPOSITION (1): *Let $(E_i, \mathcal{B}_i)_{i \in I}$ be a family of separated bornological vector spaces, let E be a vector space and, for each $i \in I$, let $u_i : E \to E_i$ be a linear map. The initial bornology on E for the maps u_i is separated if and only if, for every $x \in E$, $x \neq 0$, there exists $i \in I$ such that $u_i(x) \neq 0$.*

Proof: If the initial bornology is separated and if $x \in E$, $x \neq 0$, then the line L spanned by x is not bounded in E and hence there is an $i \in I$ for which $u_i(L)$ is not bounded in E_i. Thus $u_i(x) \neq 0$. Conversely, if the condition of the Proposition is satisfied and M is a bounded vector subspace of E, then $u_i(M)$ is a bounded subspace of E_i for every $i \in I$. Since E_i is separated, $u_i(M)$ reduces to $\{0\}$ and hence M contains no non-zero vectors.

COROLLARY: (a): *Every product of separated bornological vector spaces is separated;*

(b): *Every bornological subspace of a separated bornological vector space is separated;*

(c): *Every intersection of separated vector bornologies is separated;*

(d): *Every projective limit of separated bornological vector spaces is separated.*

The situation is not so good for final bornologies and we shall look at each case separately.

PROPOSITION (2): *Let $E = \varinjlim(E_i, v_{ji})$ be a bornological inductive limit of separated bornological vector spaces. If all the maps v_{ji} are injective, then E is separated.*

Proof: Indeed, every bounded subset of E is then a bounded subset of one of the spaces E_i, which is separated.

PROPOSITION (3): *Every bornological direct sum of separated bornological vector spaces is separated.*

Proof: The Proposition follows from Section 2:9·3 and Proposition (2). Alternatively, observe that if $E = \bigoplus_{i \in I} E_i$, with the E_i separated, then the projections $E \to E_i$ are bounded.

As for quotients, these deserve a special section.

2:11 BORNOLOGICALLY CLOSED SETS: SEPARATION OF BORNOLOGICAL QUOTIENTS

DEFINITION (1): *Let E be a bornological vector space. A set $A \subset E$ is said to be BORNOLOGICALLY CLOSED or MACKEY-CLOSED (briefly, b-CLOSED or M-CLOSED) if the conditions $(x_n)_{n \in \mathbb{N}} \subset A$ and $x_n \xrightarrow{M} x$ in E imply $x \in A$.*

REMARK (1): If E is a convex bornological space, then a set $A \subset E$ is b-closed if and only if, for every bounded disk $B \subset E$, $A \cap E_B$ is closed in E_B.

This remark is an immediate consequence of the characterisation of bornologically convergent sequences in convex bornological spaces (*cf.* Proposition (1) of Section 1:4).

REMARK (2): It can be shown that there exists a topology on E whose closed sets are exactly the b-closed subsets of E (*cf.* Exercises 1·E.9,10).

REMARK (3): Let E and F be bornological vector spaces and let $u:E \to F$ be a bounded linear map. The inverse image under u of a b-closed subset of F is b-closed in E, since $x_n \xrightarrow{M} x$ in E implies $u(x_n) \xrightarrow{M} u(x)$ in F.

PROPOSITION (1): *A bornological vector space E is separated if and only if the vector subspace $\{0\}$ is b-closed in E.*

Proof: Necessity: Suppose E to be separated, and let $A = \{0\}$ and let (x_n) be a sequence in A which converges bornologically to an element x in E. Since $x_n = 0$ for every n, this sequence also converges to 0 in E and the uniqueness of limits (Proposition (4) of Section 1:4) ensures that $x = 0$.

Sufficiency: Suppose that $\{0\}$ is b-closed in E and that (x_n) is a sequence having limits x and y in E. The sequence $x_n - x_n = 0$ converges to $x - y$, hence $x - y = 0$ and E is separated (Proposition (4) of Section 1:4).

We are now ready to give the criterion for the separation of bornological quotients.

PROPOSITION (2): *Let E be a bornological vector space and let M be a subspace of E. The quotient E/M is separated if and only if M is bornologically closed in E.*

Proof: If E/M is separated, then 0 is b-closed in E/M. If $\varphi:E \to E/M$ is the canonical map, then $M = \varphi^{-1}(0)$ is b-closed in E (Remark (2)). Conversely, suppose M b-closed in E and let H be a bounded subspace of E/M. We show that $H = \{0\}$. Let $\varphi(x) \in H$,

$x \in E$; we can find a circled bounded set $A \subset E$ such that $\mathbb{K}\varphi(x) \subset \varphi(A)$, and hence $\mathbb{K}x \subset A + M$. Thus, for every $n \in \mathbb{N}$, $nx \in A + M$ and hence there exists $x_n \in M$ such that $nx - x_n \in A$. It follows that $(x - y_n) \in (1/n)A$, where $y_n = (x_n/n) \in M$ and, therefore $y_n \xrightarrow{M} x$. Since M is b-closed, $x \in M$, which implies $\varphi(x) = 0$.

2:12 THE ASSOCIATED SEPARATED BORNOLOGICAL VECTOR SPACE

With every bornological vector space E we shall associate a separated bornological vector space $\dot{E}$ and a canonical map of E into $\dot{E}$ such that every bounded linear map of E into a separated bornological vector space factors through $\dot{E}$ in a unique way. In order to accomplish this we need the notion of *bornological closure*.

DEFINITION (1): *Let E be a bornological vector space. The* BORNOLOGICAL CLOSURE *(briefly,* b-CLOSURE *or* M-CLOSURE*) of a set $A \subset E$, denoted by $\bar{A}$, is the intersection of all bornologically closed subsets of E containing A.*

Clearly E is a b-closed set containing A. Since every intersection of b-closed sets is again b-closed, the b-closure of A is the smallest b-closed set containing A.

REMARK (1): The b-closure of a subset of E coincides with the closure for a certain topology on E (*cf.* Exercises 1·E.9,10).

PROPOSITION (1): *Let E be a bornological vector space. The b-closure of a subspace of E is again a subspace.*

Proof: First of all, it follows from the definitions that, for every b-closed set $A \subset E$ and for every $x \in E$, the set $A_x = \{z \in E;\ (x + z) \in A\}$ is b-closed in E. Let now F be a subspace of E, let $\bar{F}$ be its closure and let $x,y \in \bar{F}$. If a is any element of F, the set $\bar{F}_a$ is b-closed in E and, since it contains F, it must contain $\bar{F}$; thus $(a + y) \in \bar{F}$ for every $a \in F$. Next, the set $\bar{F}_y$ is b-closed and contains F; therefore, $\bar{F}_y \supset \bar{F}$ and hence $(x + y) \in \bar{F}$. Now let $\lambda \in \mathbb{K}$. Since the map $x \to \bar{x}$ of E into E is bounded and linear, the inverse image under this map of a b-closed set is b-closed (Remark (3) of Section 2:11). Thus the set $\{z \in E;\ \lambda z \in \bar{F}\}$ is a b-closed subset of E which contains F, hence $\bar{F}$ and, therefore, $\lambda x \in \bar{F}$ for every $x \in \bar{F}$.

REMARK (2): Note that an element of the b-closure of a set $A \subset E$ is not, in general, the bornological limit of a sequence of points of A even if A is a subspace of E (see Exercise 2·E.8). On this matter, see also Exercise 1·E.11 concerning the bornological convergence of filters.

PROPOSITION (2): *Let E be a bornological vector space, $\overline{\{0\}}$ the b-closure of $\{0\}$ in E, $\dot{E}$ the quotient $E/\overline{\{0\}}$ and $\varphi: E \to \dot{E}$ the canonical map. Then:*

(i): *$\dot{E}$ is a separated bornological vector space;*

(ii): *For every bounded linear map u of E into a separated bornological vector space G, there exists a unique bounded linear map $\dot{u}$ of $\dot{E}$ into G such that:*

$$u = \dot{u} \circ \varphi.$$

Proof: (i): This follows from the fact that $E/\overline{\{0\}}$ is separated (Proposition (2) of Section 2:11) since $\overline{\{0\}}$ is b-closed in E (Proposition (1)).

(ii): Since G is separated, 0 is b-closed in G (Proposition (1) of Section 2:11) hence $u^{-1}(0)$ is b-closed in E (Remark (3) of Section 2:11) and, containing $0 \in E$, it contains $\overline{\{0\}}$ also. Thus, if $\dot{u}$ is the map induced by u on $E/\overline{\{0\}}$, then $u = \dot{u} \circ \varphi$ and clearly $\dot{u}$ is unique.

DEFINITION (2): *The space $\dot{E}$ of Proposition (2) is called the* SEPARATED BORNOLOGICAL VECTOR SPACE ASSOCIATED WITH E *and the map $\varphi : E \to \dot{E}$ is called the* CANONICAL MAP *of* E INTO $\dot{E}$.

2:13 The Structure of a Convex Bornological Space: Comparison with the Structures of a Locally Convex Space

Let E be a convex bornological space. For two disks A and B in E such that $A \subset B$ we denote by $\pi_{BA} : E_A \to E_B$ the canonical embedding. The system (E_A, π_{BA}) is an inductive system of convex bornological spaces, since the E_A's are semi-normed spaces, and it is clear that E is the inductive limit of this system. Moreover, if E is separated, then all the spaces E_A are normed spaces and we obtain the following all-important Structure Theorem:

THEOREM (1): *Every convex bornological space E is the bornological inductive limit of a family of semi-normed spaces, and of normed spaces if E is separated.*

This Theorem shows the essential difference between the structure of a convex bornological space and that of a locally convex space. On the one hand, a convex bornology decomposes a vector space E into *internal semi-normed spaces* and reduces properties of E to those of one of its semi-normed components (e.g. bornological convergence, b-closed set). On the other hand, a locally convex topology recovers E from *external semi-normed spaces* (the spaces E_V constructed in Section 0·A.4) and obtains properties of E as intersections of properties of these external semi-normed spaces (e.g. topological convergence). In Functional Analysis it is important to be able to handle both methods of investigation.

CHAPTER III

COMPLETE BORNOLOGIES

This Chapter is devoted to *complete bornologies and complete convex bornological spaces.* These spaces are to arbitrary separated convex bornological spaces what Banach spaces are to arbitrary normed spaces. Since every separated convex bornological space is a 'union of normed spaces', a complete convex bornological space will simply be a 'union of Banach spaces' (in a precise sense).

In Section 3:3 we characterise all separated vector bornologies on a finite-dimensional vector space by showing that there is only one such bornology (up to bornological isomorphism).

In Section 3:4, with every separated bornological vector space we associate a *complete bornology on the same space* which is the closest possible to the given bornology. Such a bornology is very useful in all problems where completeness comes into play.

Finally, Section 3:5 introduces the notion of *bornological completeness* for a locally convex space, which, although less restrictive than the usual notions of completeness, is generally sufficient for the needs of Functional Analysis.

Supplementary results and counter-examples can be found in the Exercises.

3:1 COMPLETANT BOUNDED DISKS

DEFINITION (1): *Let E be a vector space. A disk $A \subset E$ is called a* COMPLETANT DISK *if the space E_A spanned by A and semi-normed by the gauge of A is a Banach space.*

In order to give a general example of a completant bounded disk, let us recall that a subset A of a separated topological vector space E is said to be *SEQUENTIALLY COMPLETE* if every Cauch sequence of elements of A converges to an element of A. The set A is then *SEQUENTIALLY CLOSED,* i.e. it contains the limit of ever sequence of elements of A which converges in E. Clearly, if A is sequentially complete, so is λA for every $\lambda \in \mathbb{K}$.

PROPOSITION (1): *Let E be a separated topological vector space. Every bounded and sequentially complete disk $A \subset E$ is completant.*

Proof: We have to show that E_A is a Banach space. Since E is separated, E_A is normed. Let (x_n) be a Cauchy sequence in E_A. Since the canonical embedding $E_A \to E$ is linear and continuous (A is bounded in E), (x_n) is a Cauchy sequence in E. Now (x_n) is bounded in E_A, hence contained in λA for some $\lambda \in \mathbb{K}$ and, since λA is sequentially complete, (x_n) converges in E to a point $x \in \lambda A \subset E_A$. It remains to show that (x_n) converges to x in the norm of E_A. This will be a consequence of the fact that A is sequentially closed in E, by virtue of the following general argument. Since (x_n) is a Cauchy sequence in E_A, given $\varepsilon > 0$ there exists an integer $N(\varepsilon)$ such that $(x_m - x_n) \in \varepsilon A$ for $m,n \geqslant N(\varepsilon)$. We fix $m \geqslant N(\varepsilon)$ and let $n \to +\infty$; then $(x_m - x_n) \to (x_m - x)$ in E and hence $(x_m - x) \in \varepsilon A$ for εA is sequentially closed. Thus $(x_m - x) \in \varepsilon A$ for every $m \geqslant N(\varepsilon)$, i.e. $x_m \to x$ in E_A.

COROLLARY: *Let E be a vector space and let A be a disk in E which is compact for some separated vector topology on E. Then A is completant.*

Proof: By Proposition (1) it suffices to show that every compact disk A of a separated topological vector space E is sequentially complete. Let then (x_n) be a Cauchy sequence in A and denote by F_n the closure of the set $\{x_p : p \geqslant n\}$. The compactness of A ensures that $\bigcap_{n=1}^{\infty} F_n \neq \emptyset$. Let $x \in \bigcap_{n=1}^{\infty} F_n$; we show that (x_n) converges to x in E. In fact, (x_n) being a Cauchy sequence, for every circled neighbourhood V of 0 in E, there exists an integer N such that $(x_p - x_q) \in V$ whenever $p,q \geqslant N$. Since $x \in F_n$, there is a $p \geqslant N$ such that $(x_p - x) \in V$. Then for every $q \geqslant N$ we have:

$$x_q - x = (x_q - x_p) + (x_p - x) \in V + V,$$

and the assertion follows.

PROPOSITION (2): *Let E, F be separated bornological vector spaces and let $u: E \to F$ be a bounded linear map. If A is a completant bounded disk in E, then $u(A)$ is a completant bounded disk in F.*

Proof: Indeed, $F_{u(A)}$ is isometric to a separated quotient of the Banach space E_A (Proposition (3) of Section 0·A.3) and hence is a Banach space.

PROPOSITION (3): *Let I be a non-empty set of indices and let $(E_i)_{i \in I}$ be a family of vector spaces. For every $i \in I$, let A_i be a completant disk in E_i. If $E = \prod_{i \in I} E_i$ and $A = \prod_{i \in I} A_i$, then A is a completant disk in E.*

Proof: By virtue of Proposition (2) of Section 0·A.4, $E_A = \{x = (x_i);\ \sup_{i \in I} p_{A_i}(x_i) < +\infty\}$, p_{A_i} denoting the gauge of A_i. Furthermore, $p_A(x) = \sup_{i \in I} p_{A_i}(x_i)$. If, then, $(x^{(n)})$ is a Cauchy sequence in E_A, for every $i \in I$ $(x_i^{(n)})$ is a Cauchy sequence in $(E_i)_{A_i}$, hence it converges to an element $w_i \in (E_i)_{A_i}$, since $(E_i)_{A_i}$ is complete. It is now immediate that $x = (x_i) \in E_A$ and that the sequence $(x^{(n)})$ converges to x in E_A.

3:2 COMPLETE CONVEX BORNOLOGICAL SPACES

3:2·1

DEFINITION (1): *A convex bornology on a vector space is called a* COMPLETE CONVEX BORNOLOGY *if it has a base consisting of completant disks. A convex bornological space is called a* COMPLETE CONVEX BORNOLOGICAL SPACE *if its bornology is complete.*

Any such space is separated, by definition.

3:2·2 Structure of Complete Convex Bornological Spaces

Let E be a complete convex bornological space and let $\mathcal{B}$ be a base for its bornology consisting of completant disks. For every $A \in \mathcal{B}$, E_A is a Banach space and, as in Section 2:13, one proves that E is the inductive limit of the Banach spaces E_A. Conversely it is clear that every bornological inductive limit of a bornological inductive system (E_i, u_{ji}) of Banach spaces, with all the maps u_{ji} injective, is a complete convex bornological space. Thus we have established the following result: A convex bornological space is complete if and only if it is the bornological inductive limit of Banach spaces with injective maps. Hence, *complete convex bornological spaces are to separated convex bornological spaces what Banch spaces are to normed spaces*, and it is precisely this fact that motivates their interest.

3:2·3 Stability Properties

Complete convex bornological spaces have good stability properties, as we shall show presently.

PROPOSITION (1): *Let E be a separated convex bornological space and let F be a bornological subspace of E. Then:*

(i): *If F is complete, it is b-closed in E;*

(ii): *If E is complete and F is b-closed, then F is complete.*

Proof: (i): Let (x_n) be a sequence in F which converges bornologically to x in E; there exists a bounded disk $A \subset E$ such that $x_n \to x$ in E_A. Since $A \cap F$ is bounded in F and F is complete, we can find a completant bounded disk $B \subset F$ such that $A \cap F \subset B$. Now (x_n) is a Cauchy sequence in E_A and is contained in F, whence it is a Cauchy sequence in F_B and, therefore, it converges to an element $y \in F_B$. But the embedding $F_B \to F \to E$ is bounded, hence (x_n) converges to y in E and we must have $y = x$, for E is separated.

(ii): Let $\mathcal{B}$ be a base for the bornology of E consisting of completant disks. It is enough to show that, for every $B \in \mathcal{B}$, the set $A = B \cap F$ is completant. Let then (x_n) be a Cauchy sequence in $F_A = E_B \cap F$. Since (x_n) is also a Cauchy sequence in the Banach space E_B, it must converge to a point $x \in E_B$. But F_A, equipped with its norm (the gauge of A), is a closed subspace of E_B, hence $x \in F_A$ and $x_n \to x$ in F_A.

PROPOSITION (2): *If E is a complete convex bornological space and F is a b-closed subspace of E, then the quotient E/F is complete.*

Proof: Let $\mathcal{B}$ be a base for the bornology of E consisting of completant disks. If $\varphi : E \to E/F$ is the canonical map, then $\varphi(\mathcal{B})$ is a base for the bornology of E/F. Since E/F is separated (Proposition (2) of Section 2:11) for every $A \in \mathcal{B}$, $\varphi(A)$ is a completant disk in E/F (Proposition (2) of Section 3:1), whence E/F is complete.

PROPOSITION (3): *Every product of complete convex bornological spaces is complete.*

Proof: This follows from Proposition (3) of Section 3:1.

PROPOSITION (4): *Let $(E_i, u_{ji})_{i \in I}$ be an inductive system of complete convex bornological spaces (i.e., E_i is complete for every $i \in I$), and let $E = \varinjlim_{i \in I} E_i$. Then E is complete if and only if E is separated.*

Proof: Since every complete space is separated, only the sufficiency needs proving. Assume, then, E to be separated and let u_i be the canonical embedding of E_i into E. A base for the bornology of E is formed by the disks $u_i(A)$ where A runs through all the completant disks of a base for E_i and i runs through I. But E is separated, hence each $u_i(A)$ is completant (Proposition (3) of Section 3:1) and consequently E is complete.

COROLLARY (1): *With the notation of Proposition (4), if the maps u_{ji} are injective, then E is complete.*

For E is then necessarily separated (Proposition (2) of Section 2:10).

COROLLARY (2): *Every bornological direct sum of complete convex bornological spaces is complete.*

Proof: Let $(E_i)_{i \in I}$ be a family of complete convex bornological spaces and let $E = \bigoplus_{i \in I} E_i$ be their bornological direct sum. Denote by $\mathfrak{F}(I)$ the set of finite subsets of I, directed under inclusion. For every $J \in \mathfrak{F}(I)$ let $E_J = \bigoplus_{i \in I} E_i$. The space E_J is bornologically isomorphic to the product $\prod_{i \in J} E_i$, whence is complete (Proposition (3)). If $J \subset J'$, denote by $u_{J'J}$ the canonical embedding of E_J into $E_{J'}$. Then E is the bornological inductive limit of the spaces E_J and the assertion follows from Corollary (1).

3:3 Separated Bornological Vector Spaces of Finite Dimension

In this Section we shall show that, up to a bornological isomorphism, $\mathbb{K}^n$, endowed with its canonical bornology, is the only separated bornological vector space of dimension n (n a positive integer). This is the analogue of a well known result for separated topological vector spaces (see Bourbaki [3]).

LEMMA (1): *Every separated bornological vector space E of dimension 1 is bornologically isomorphic to the scalar field $\mathbb{K}$ equipped with its canonical bornology.*

Proof: Let $a \in E$, $a \neq 0$. The linear map $u: \lambda a \to \lambda$ of E into $\mathbb{K}$ is an algebraic isomorphism whose inverse $\lambda \to \lambda a$ is bounded. In order to show that u is a bornological isomorphism, it is then enough to show that u is bounded. Suppose not, and let B be a circled bounded subset of E such that $u(B)$ is not bounded. Now the only circled unbounded subset of $\mathbb{K}$ is $\mathbb{K}$ itself (direct verification), whence $u(B) = \mathbb{K}$ and, consequently, $B = E$. However, this is impossible, since E is separated and B is bounded.

LEMMA (2): *Let E be a bornological vector space. Every one-dimensional subspace D of E which is the algebraic complement of a b-closed hyperplane $H \subset E$, is also a bornological complement of H.*

Proof: $\{0\}$ is b-closed in D, since it is the intersection of D and the b-closed hyperplane H. It follows that D is separated for the bornology induced by E (Proposition (1) of Section 2:11). Since E/H is separated for the quotient bornology (Proposition (2) of Section 2:11), the canonical algebraic isomorphism between D and E/H is also a bornological isomorphism (Lemma (1)), hence the Lemma.

We now have:

THEOREM (1): *Every separated bornological vector space E of finite dimension n is bornologically isomorphic to $\mathbb{K}^n$, where $\mathbb{K}$ is the scalar field endowed with its canonical bornology.*

Proof: The Theorem has already been proved for $n = 1$ (Lemma (1)). Hence we shall assume that the statement of the Theorem to be true for $n - 1$ and prove it to be true for n. Every hyperplane in E is a separated bornological vector space of dimension $n - 1$ for the induced bornology, and hence is isomorphic to $\mathbb{K}^{n-1}$ by assumption. Now $\mathbb{K}^{n-1}$ is complete, since its unit ball for any norm is compact, whence completant (Corollary to Proposition

(1) of Section 3:1). It follows that H is complete and, by Proposition (1) of Section 3:2, b-closed in E. Lemma (2) now implies that any algebraic complement D of H is also a bornological complement, which means that E is bornologically isomorphic to $H \oplus D$, hence to $\mathbb{K}^{n-1} \oplus \mathbb{K} = \mathbb{K}^n$ and the Theorem is proved.

3:4 THE COMPLETE BORNOLOGY ASSOCIATED WITH A SEPARATED VECTOR BORNOLOGY

With every separated bornological vector space E one can associate a complete convex bornological space E_0, *algebraically identical to* E, with the following 'co-universal' property: Every bounded linear map of a complete convex bornological space F into E is also a bounded linear map of F into E_0. In all those questions in which completeness plays an essential part, the bornology of E_0 will often take the place of that of E, with the considerable advantage that we shall be working with a complete space, whilst remaining in the same vector space.

LEMMA (1): *The family $\mathcal{B}$ of all completant bounded disks of a separated bornological vector space E is a base for a complete bornology on E.*

Proof: We have to show that $\mathcal{B}$ is a covering of E which is stable under the formation of vector sums and scalar multiples. First of all, $\mathcal{B}$ covers E: in fact, every point of E lies in a finite-dimensional bounded disk and hence in a completant bounded disk (Theorem (1) of Section 3:3). Next, the sum $A + B$ of two completant bounded disks A and B is a completant bounded disk, since $E_{(A+B)}$ is isometric to a separated quotient of the Banach space $E_A \times E_B$ (Proposition (1) of Section 0·A.4). Finally, it is clear that scalar multiples of members of $\mathcal{B}$ belong to $\mathcal{B}$, and the Lemma is proved.

PROPOSITION (1): *Let E be a separated bornological vector space and denote by E_0 the vector space E furnished with the bornology having the family of all completant bounded disks of E as a base. Let i be the canonical embedding of E_0 into E: i is linear and bounded. Then the pair (E_0, i) has the following properties:*

(i): *E_0 is a complete convex bornological space;*

(ii): *For every bounded linear map u of a complete convex bornological space F into E, there exists a unique bounded linear map u_0 of F into E_0 such that:*

$$u = i \circ u_0.$$

Proof: (i): This is obvious, by definition of E_0 (Lemma (1)). For (ii), note that the image under u of a completant bounded disk of F is bounded in E and completant (Proposition (2) of Section 3:1), hence bounded in E_0. Let, then, u_0 be the map u regarded as a map from F to E_0; it is clear that (ii) holds.

DEFINITION (1): *With the notation of Proposition* (1), E_0 *is called the* COMPLETE CONVEX BORNOLOGICAL SPACE ASSOCIATED WITH E.

3:5 BORNOLOGICALLY COMPLETE TOPOLOGICAL VECTOR SPACES

The purpose of this Section is to give a simple criterion, in terms of convergence of sequences, for the von Neumann bornology of a separated locally convex space to be complete.

DEFINITION (1): *A separated locally convex space E is called* BORNOLOGICALLY COMPLETE *if its von Neumann bornology is complete.*

DEFINITION (2): *Let E be a separated convex bornological space. A sequence (x_n) in E is said to be a* BORNOLOGICAL CAUCHY SEQUENCE *(or a* MACKEY-CAUCHY SEQUENCE*) in E if there exists a bounded disk $B \subset E$ such that (x_n) is a Cauchy sequence in E_B.*

If E is a separated locally convex space we shall speak, with abuse of language, of Mackey-Cauchy sequences in E to mean Mackey Cauchy sequences in the space E equipped with its von Neumann bornology.

PROPOSITION (1): *A separated locally convex space E is bornologically complete if (and only if) every Mackey-Cauchy sequence in E is topologically convergent.*

Proof: The condition is clearly necessary for a general convex bornological space: if such a space is complete, a Mackey-Cauchy sequence is obviously bornologically convergent, since in Definition (2) B can be chosen to be completant. It is the sufficienc that is peculiar to a particular class of convex bornological spaces containing the von Neumann bornologies of locally convex spaces (see Exercises 3·E.2,3). Thus let us show that the condition of the Proposition is sufficient. Let $\mathcal{B}$ be a base for the von Neumann bornology of E consisting of *closed* disks. We claim that every member of $\mathcal{B}$ is completant. In fact, let $A \in \mathcal{B}$ and let (x_n) be a Cauchy sequence in E_A. Then (x_n) is a Mackey-Cauchy sequence in E and, by assumption, (x_n) converges topologically to some $x \in E$. It follows, since A is closed, that $x \in E_A$ and that (x_n) converges to x in E_A (*cf.* end of proof of Proposition (1) of Section 3:1, which gives a general argument).

COROLLARY: *A separated locally convex space E, in which every Cauchy sequence converges, is bornologically complete.*

Proof: This follows from Proposition (1) and the fact that every Mackey-Cauchy sequence in E is a Cauchy sequence.

REMARK: The above Corollary shows that, for topological vector spaces, bornological completeness is much weaker than completene Nevertheless, for a great many problems bornological completenes turns out to be enough.

CHAPTER IV

'TOPOLOGY—BORNOLOGY': INTERNAL DUALITY

There is a triple duality between locally convex spaces and convex bornological spaces. First of all, we have the duality within the same given vector space E, which we call *internal duality*. This duality associates, in a natural way, with every locally convex topology on E a canonical bornology and with every convex bornology on E a canonical topology, and investigates their 'functorial' interplay. This leads to the notions of *bornological topology* and *topological bornology*. The importance of bornological topologies is made clear in Section 4:2: they make bounded linear maps continuous. The analysis of the internal duality also leads, quite naturally, to the notion of a *completely bornological topology* (Section 4:3): under this topology all linear maps which are bounded on completant bounded disks (*a fortiori* on complete bounded disks) are continuous. In the literature such a topology is also called *ultra-bornological*, but this terminology does not make sufficiently precise in what respects this topology differs from a bornological topology. Finally, Section 4:4 is devoted to the Closed Graph Theorem, where completely bornological spaces play the principal rôle, whilst several simple counter-examples to the theory expounded in this Chapter are given in the Exercises.

The two other aspects of the duality between topology and bornology concern the external duality, which is treated in Chapters V,VI.

4:1 COMPATIBLE TOPOLOGIES AND BORNOLOGIES

4:1·1 Definition of Compatibility

Let E be a vector space and let $\mathcal{B}$ (resp. $\mathcal{T}$) be a bornology (resp. a vector topology) on E; $\mathcal{B}$ need not be a vector bornology. We say that $\mathcal{B}$ and $\mathcal{T}$ are *COMPATIBLE* if $\mathcal{B}$ is finer than the von Neumann bornology of $(E,\mathcal{T})$. This simply means that the identity $(E,\mathcal{B}) \to (E,\mathcal{T})$ is bounded.

4:1·2 The Space ^tE

A subset of a bornological vector space E is called a *BORNIVOROUS SUBSET* if it absorbs every bounded subset of E. Several properties of bornivorous sets are given in Exercise 1·E.8.

Let E be a convex bornological space and let $\mathcal{V}$ be the family of all bornivorous disks in E. We are going to show that *$\mathcal{V}$ is a base of neighbourhoods of 0 for the finest locally convex topology on E compatible with the bornology of E.* The members of $\mathcal{V}$ are absorbent and, by definition, convex and circled. It is clear that $\mathcal{V}$ is stable under finite intersections and homothetic transformations, hence $\mathcal{V}$ is a base of neighbourhoods of 0 for a locally convex topology $\mathcal{T}$ on E. Every bounded subset of E, being absorbed by any member of $\mathcal{V}$, is bounded in $(E,\mathcal{T})$ in the von Neumann sense. $\mathcal{T}'$ is a locally convex topology on E which is compatible with the bornology $\mathcal{B}$ of E, then $\mathcal{T}'$ has a base of neighbourhoods of zero consisting of bornivorous disks of E and hence is coarser than $\mathcal{T}$.

The topology $\mathcal{T}$ just defined is called the *LOCALLY CONVEX TOPOLOGY ASSOCIATED WITH THE BORNOLOGY of E* and the space E, *endowed with this topology, is denoted by ^tE or $\mathbb{T}E$.*

4:1·3 The Space ^bE

Let $(E,\mathcal{T})$ be a locally convex space. *There exists on E a coarsest convex bornology compatible with $\mathcal{T}$: it is precisely the von Neumann bornology of* $(E,\mathcal{T})$, as follows from the definitions. Endowed with such a bornology, the space E *will be denoted by* ^bE *or* $\mathbb{B}E$.

4:1·4 The Topological Bornology

If E is a convex bornological space, the bornology of ^{bt}E is always coarser than the original bornology of E (*cf.* Exercise 4·E.1) since, by definition of ^tE, each bounded subset of E is absorbed by every neighbourhood of 0 in ^tE. The bornology of ^{bt}E is called the *WEAK BORNOLOGY* of E. The following Proposition gives a criterion for this bornology to agree with the original bornology of E.

PROPOSITION (1): *Let E be a convex bornological space. Then:*

$$E = {}^{bt}E,$$

if and only if the bornology of E is the von Neumann bornology of a locally convex topology on E.

Proof: The necessity is obvious, since then the bornology of E is the von Neumann bornology of ^tE. For the sufficiency, let $\mathcal{T}$ be a locally convex topology on E and denote by F the locally convex space $(E,\mathcal{T})$. By hypothesis we have $E = {}^bF$ and hence ^{bt}E ^{btb}F. The assertion will then be a consequence of the following general Lemma.

LEMMA (1): *For every locally convex space F we have the bornological identity:*

$$^{b}F = {}^{btb}F.$$

Proof: The Lemma expresses the fact that F and ^{tb}F have the same bounded sets. First of all, since the identity $^{tb}F \to F$ is continuous, the identity $^{btb}F \to {}^{b}F$ is bounded (direct verification). Conversely, let B be a bounded subset of ^{b}F. By definition of ^{tb}F, B is absorbed by every neighbourhood of 0 in ^{tb}F, hence is bounded in ^{btb}F and the Lemma follows.

The Proposition (1) is now proved, since:

$$E = {}^{b}F = {}^{btb}F = {}^{bt}E.$$

The following definition finds its justification in Proposition (1):

DEFINITION (1): *Let E be a convex bornological space. We say that the bornology of E is a topological bornology, or that E is a topological convex bornological space, if the following bornological identity holds:*

$$E = {}^{bt}E.$$

In the light of Definition (1), the above Proposition (1) can then be formulated by saying that a convex bornology is topological if and only if it is the von Neumann bornology of a locally convex topology. By virtue of Lemma (1), the bornology of ^{bt}E, with E a convex bornological space, is always a topological bornology.

4:1·5 The Bornological Topology

If E is a locally convex space, the topology of ^{tb}E is always finer than the original topology of E and, in general, strictly finer (*cf.* Exercise 4·E.2). A necessary and sufficient condition for these two topologies to agree is given by the following Proposition.

PROPOSITION (2): *Let E be a locally convex space. Then:*

$$E = {}^{tb}E,$$

if and only if the topology of E is the locally convex topology associated with a convex bornology on E.

Proof: The necessity is obvious, since then the topology of E is the locally convex topology associated with the bornology of ^{b}E. For the sufficiency, let $\mathcal{B}$ be a convex bornology on E and denote by F the convex bornological space $(E,\mathcal{B})$. By hypothesis we have $E = {}^{t}F$ and hence $^{tb}E = {}^{tbt}F$. The assertion will then be a consequence of the following general Lemma.

LEMMA (2): *For every convex bornological space we have the topological identity:*

$$^{t}F = {}^{tbt}F.$$

Proof: Since the identity $F \to {}^{bt}F$ is bounded, the identity ${}^{t}F \to {}^{tbt}F$ is continuous (direct verification). Conversely, let V be a neighbourhood of 0 in ${}^{t}F$, which may be assumed to be a bornivorous disk of F. We have to show that V is a neighbourhood of 0 in ${}^{tbt}F$, i.e. a disk which absorbs the bounded subsets of ${}^{bt}F$. Now the bounded subsets of ${}^{bt}F$ are exactly those subsets of F that are absorbed by every neighbourhood of 0 in ${}^{t}F$, whence by every bornivorous disk of F, and, therefore, V is a neighbourhood of 0 in ${}^{tbt}F$.

Proposition (2) is now proved, since:

$$E = {}^{t}F = {}^{tbt}F = {}^{tb}E.$$

In the light of Proposition (2) we give the following definition:

DEFINITION (2): *Let E be a locally convex space. We say that the topology of E is a* BORNOLOGICAL TOPOLOGY, *or that E is a* BORNOLOGICAL LOCALLY CONVEX SPACE, *if the following topological identity holds:*

$$E = {}^{tb}E.$$

Proposition (2) can now be formulated by saying that a locally convex topology is bornological if and only if it is the locally convex topology associated with a convex bornology. By Lemma (2) the topology of ${}^{t}F$, with F a convex bornological space, is always a bornological topology.

A simple example of a locally convex topology that is not born ological can be found in Exercise 4·E.2. Here we give an important example of a bornological topology.

PROPOSITION (3): *Every metrizable locally convex topology is bornological.*

Proof: We have now to show that if E is a metrizable locally convex space, then $E = {}^{tb}E$. Since the topology of ${}^{tb}E$ is always finer than that of E, it will suffice to prove that the identity $E \to {}^{tb}E$ is continuous, i.e. that every bornivorous disk of ${}^{b}E$ is a neighbourhood of 0 in E. But such a disk absorbs all sequences that converge to 0, since these are bounded in ${}^{b}E$, whence is a neighbourhood of 0 by virtue of the following Lemma.

LEMMA (3): *In a metrizable topological vector space E, every circled set which absorbs all sequences converging to* 0 *is a neighbourhood of* 0.

Proof: Let (V_n) be a decreasing base of neighbourhoods of 0 in E and let P be a circled subset of E which absorbs all sequences that converge to 0. If P is not a neighbourhood of 0, then it contains no set of the form $(1/n)V_n$ and hence there exists a sequence (x_n) in E such that $x_n \in V_n$ and $x_n \notin nP$. The sequence (x_n) then converges to 0, yet is not absorbed by P, contradicting the hypothesis on P.

For the permanence properties of bornological topologies the reader is referred to Exercise 4•E.4.

4:2 CHARACTERISATION OF BORNOLOGICAL TOPOLOGIES

4:2˙1 Formulation of the Problem

Let E,F be locally convex spaces and let $u:E \to F$ be a linear map. We have already made use, in Section 4:1, of the following fact: if u is continuous, then it is bounded (for the von Neumann bornologies of E and F). In fact, let A be a bounded subset of ${}^bE$ and let V be a neighbourhood of 0 in F. Since u is continuous, $u^{-1}(V)$ is a neighbourhood of 0 in E and hence absorbs A. Thus $u(u^{-1}(V)) = V$ absorbs A and, consequently, $u(A)$ is bounded in ${}^bF$.

The converse of the above assertion is generally false; in other words, if the linear map u is bounded, it does not follow that u is continuous, even if F is the scalar field (Exercise 4•E.2), and it is an important problem to know for which locally convex spaces the continuity of a linear map follows from its boundedness. The importance of this problem rests on the following two reasons: the first is that bounded linear maps are encountered very frequently; the second, that in almost all cases it is much easier to show the boundedness of a linear map than its continuity.

4:2˙2

The result we are going to establish asserts that the locally convex topologies on E for which every bounded linear map of E into any locally convex space is continuous are exactly the bornological topologies. Precisely, we have:

PROPOSITION (1): *Let E be a locally convex space. The following assertions are equivalent:*

(i): *E is a bornological locally convex space;*

(ii): *Every bounded linear map of E into an arbitrary locally convex space is continuous.*

Proof: (i) => (ii): Let u be a bounded linear map of E into a locally convex space F. Then for every disked neighbourhood V of 0 in F, $u^{-1}(V)$ is a bornivorous disk in E, hence a neighbourhood of 0, since $E = {}^{tb}E$.

(ii) => (i): Let D be a bornivorous disk in ${}^bE$ with gauge p_D; p_D is a semi-norm on E. Denote by E_D the space E furnished with the semi-norm p_D. The identity $u:E \to E_D$ is bounded, since D is bornivorous, hence continuous and, therefore, $u^{-1}(D) = D$ is a neighbourhood of 0 in E. From this the topological identity ${}^{tb}E = E$ follows, since the identity ${}^{tb}E \to E$ is always continuous.

4:2·3

In order to give other characterisations of the bornological topology we need the following Lemma.

LEMMA (1): *Let E and F be bornological vector spaces and suppose that one of the following conditions is satisfied:*

(i): *The bornology of F has a countable base;*

(ii): *The bornology of F is the von Neumann bornology of a vector topology on E.*

Let u be a linear map of E into F. If u maps every bornologically convergent sequence in E onto a bounded sequence in F, then u is bounded.

Proof: (i): *The bornology of F has a countable base:* Let (B_n) be a base for the bornology of F, consisting of an increasing sequence of circled bounded sets. If the map u is not bounded, there exists a bounded set $A \subset E$ such that, for every $n \in \mathbb{N}$, $u(A) \not\subset nB_n$; hence, there is a sequence (a_n) in A such that $u(a_n) \notin nB_n$. The sequence $((1/n)a_n)$ converges bornologically to 0 in E, but its image under u in F is unbounded, otherwise there would be an $n_0 \in \mathbb{N}$ for which $u((1/n)a_n) \subset B_{n_0}$ for all $n \in \mathbb{N}$, contradicting the fact that $u(a_{n_0}) \notin n_0B_{n_0}$.

(ii): *The bornology of F is a von Neumann bornology:* The proof is similar to (i): assuming u to be unbounded, there must be a bounded set $A \subset E$ such that $u(A)$ is not absorbed by some neighbourhood V of 0 for the vector topology considered on F. It follows that, for every $n \in \mathbb{N}$, $u(A) \not\subset n^2V$ and hence that A contains a sequence (a_n) such that $u(a_n) \notin n^2V$. Now the sequence $((1/n)a_n)$ converges bornologically to 0 in E but its image in F is unbounded, since it is not absorbed by V.

From Lemma (1) we deduce the following Theorem, which gathers together the most useful characterisations of the bornological topology.

THEOREM (1): *Let E be a locally convex space. The following assertions are equivalent:*

(i): *E is a bornological locally convex space;*

(ii): *Every bounded linear map of E into a locally convex space F is continuous;*

(iii): *Every linear map of E into a locally convex space F, which is bounded on each compact subset of E, is continuous;*

(iv): *Every linear map of E into a locally convex space F, which is bounded on each sequence that converges to 0 in E, is continuous.*

(v): *Every linear map of E into a locally convex space F, which is bounded on each sequence that converges bornologically to 0 in* ^bE, *is continuous.*

Proof: It is evident that $(k) \Rightarrow (k-1)$ for (k) = (ii), (iii), (iv), (v). In fact: every sequence which converges bornologically to 0 also converges topologically to 0; every sequence which converges topologically to 0 is relatively compact; every compact subset of E is bounded in ^bE, and, finally, (ii) => (i) by Proposition (1). Thus it suffices to prove the implication (i) => (v). But every linear map of E into an arbitrary locally convex space F, which is bounded on each sequence that converges bornologically to 0 in ^bE, is also bounded as a map from ^bE to ^bF (Lemma (i)(ii)), whence it is continuous by (i).

REMARK (1): Theorem (1) gives 'external' characterisations of the bornological topology, since the auxiliary space F, other than E itself, appears in its statement. However, all such characterisations can be formulated 'internally', in terms of the space E alone, as shown in Exercise 4·E.6.

4:3 COMPLETELY BORNOLOGICAL SPACES

4:3·1 Formulation of the Problem

Let E,F be locally convex spaces and let $u:E \to F$ be a linear map. We know from Theorem (1) of Section 4:2 that if E is bornological, then u is continuous if it is bounded on each compact subset of E. Suppose we only know that u is bounded on each *compact convex* subset of E; can we deduce the continuity of u when E is bornological? The answer is negative in general, even if E is a normed space (*cf.* Exercise 4·E.3) and the problem arises of how to characterise those (necessarily bornological) locally convex spaces for which the above question has a positive answer. This is a very important problem in Functional Analysis, since it is the key to the 'General Closed Graph Theorem' (*cf.* Section 4:4 below). In this Section we shall characterise all those locally convex spaces for which the above question can be answered in the affirmative: they are the 'completely bornological spaces', which we are now going to define.

4:3·2 Definition and Examples of Completely Bornological Spaces

In order to understand the definition of completely bornological spaces let us recall that a locally convex space E is bornological if and only if there exists a convex bornological space E_1 such that $E = {}^tE_1$ (Proposition (2) of Section 4:1).

DEFINITION (1): *A separated locally convex space E is called* COMPLETELY BORNOLOGICAL *(or* ULTRA-BORNOLOGICAL*) if there exists a complete convex bornological space* E_1 *such that* $E = {}^tE_1$ *algebraically and topologically. If this is the case, the topology of E is then called a* COMPLETELY BORNOLOGICAL TOPOLOGY.

Trivially, every completely bornological topology is bornological. For an example of a space which is normed (hence bornological) and not completely bornological, see Exercise 4•E.3.

The following Proposition, an immediate consequence of the definitions, gives a sufficient condition for a bornological topology to be completely bornological.

PROPOSITION (1): *Every separated locally convex space which is bornological and bornologically complete (Section* 3:5*) is completely bornological.*

COROLLARY: *Every Fréchet space is completely bornological.*

In fact, a Fréchet space is bornological (Proposition (3) of Section 4:1) and bornologically complete (Corollary to Proposition (1) of Section 3:5).

Other important examples and constructions of completely bornological spaces can be found in Exercise 4•E.4 and in the following Chapter V. Furthermore, the most common spaces that occur in practice are completely bornological.

4:3·3 Characterisations of Completely Bornological Spaces

Let E be a separated locally convex space and denote by E_0 the complete convex bornological space associated with ${}^bE$ (Definition (1) of Section 3:4); a base for the bornology of E_0 consists of all completant bounded disks of ${}^bE$. We have:

THEOREM (1): *The following assertions are equivalent:*

(i): *E is completely bornological;*

(ii): *Every linear map of E into a locally convex space F, which is bounded on each completant bounded disk of ${}^bE$, is continuous;*

(iii): *Every linear map of E into a locally convex space F, which is bounded on each compact disk of E, is continuous;*

(iv): *Every linear map of E into a locally convex space F, which is bounded on each sequence that converges bornologically to 0 in E_0, is continuous.*

Proof: First of all, observe that if u is a linear map of E into a locally convex space F and u is bounded on each sequence that converges bornologically to 0 in E_0, then u is bounded from E_0 into F (Lemma (1) of Section 4:2). It follows that assertions (ii) and (iv) are equivalent and it remains to prove the implications (i) => (ii) => (iii) => (i).

(i) => (ii): Let $u:E \to F$ be a linear map as in (ii). Since E is completely bornological, $E = {}^tE_1$ where E_1 is a complete convex bornological space. It is enough to show that u is bounded from E_1 to ${}^bF$, since then u is continuous from ${}^tE_1 = E$ to ${}^{tb}F$ and, *a fortiori*, to F. Let B be a bounded disk in E_1, which may be assumed to be completant; B is bounded in ${}^{bt}E_1 = {}^bE$ and, since it is completant, it is also bounded in E_0. Hence u is bounded on B by assumption and the assertion follows.

(ii) => (iii): Since (ii) and (iv) are equivalent, we show that (iv) implies (iii). Let the linear map $u:E \to F$ be bounded on every compact disk of E and let (x_n) be a sequence which converges bornologically to 0 in E_0. There exists a completant bounded disk $B \subset E$ such that (x_n) converges to 0 in the Banach space E_B. Since the closed disked hull of the compact set $A = (x_n) \cup \{0\}$ is a compact disk in E_B (Example (10) of Section 1:3), u is bounded on A by assumption and, consequently, u is continuous by (iv).

(iii) => (i): Denote by $\mathcal{K}$ the bornology of compact disks of E (Example (6) of Section 1:3); $\mathcal{K}$ is a complete bornology. Putting $E_1 = (E,\mathcal{K})$, we show that $E = {}^tE_1$ (topologically). By (iii) the identity $E \to {}^tE_1$ is continuous, since it is bounded on each member of $\mathcal{K}$ (in fact, each member of $\mathcal{K}$ is bounded in E_1, whence in ${}^{bt}E_1$). Conversely, if V is a disked neighbourhood of 0 in E, then V is a bornivorous disk in E, *a fortiori*, in E_1, which means that V absorbs every compact disk in E. It follows that V is a neighbourhood of 0 in ${}^tE_1$ and hence the identity ${}^tE_1 \to E$ is continuous. Therefore, the topology of ${}^tE_1$ is the same as the given topology of E and (i) follows. The Theorem is now completely proved.

4:4 THE CLOSED GRAPH THEOREM

4:4·1 Formulation of the Problem

The following situation occurs very frequently in Functional Analysis. A bounded (resp. continuous) linear map $u:E \to F$ is given between two convex bornological spaces (resp. locally convex spaces) E and F; u takes its values in a subspace G of F which is equipped with a finer convex bornology (resp. finer locally convex topology) than that induced by F. When can we say that the map u is bounded (resp. continuous) as a map of E into G? The Closed Graph Theorem provides a very general answer to this question.

4:4·2 The Graph of a Map

Let X and Y be two sets and let u be a map from X to Y. The *GRAPH OF A MAP* u is the set of all pairs $(x,y) \in X \times Y$ such that $y = u(x)$. If X and Y are vector spaces and u is linear, the graph of u is a vector subspace of $X \times Y$. If X and Y are separated topological vector spaces (resp. separated bornological vector spaces) and if u is linear and continuous (resp. linear and bounded), then the graph of u is closed (resp. b-closed) in the space $X \times Y$ endowed with the product topology (resp. product bornology). Let us prove the assertion, for example, if X and Y are separated bornological vector spaces. Denote by A the graph of u and let $(x_n, u(x_n))$ be a sequence in A which converges bornologically to (x,y) in $X \times Y$. Then $x_n \xrightarrow{M} x$ in X and $u(x_n) \xrightarrow{M} y$ in Y. Since u is bounded, the sequence $(u(x_n))$ converges bornologically to $u(x)$ and, since Y is separated, we must have $y = u(x)$. Therefore $(x,y) \in A$ and A is b-closed in $X \times Y$.

The Closed Graph Theorem is, in a sense, the converse of the above assertion. It essentially states that if $u:X \to Y$ is linear and has a closed (resp. b-closed) graph, then u is continuous (resp. bounded) provided X and Y belong to suitable classes of topological vector spaces (resp. bornological vector spaces). In this Section we shall prove a General Closed Graph Theorem for locally convex spaces and a General b-Closed Graph Theorem for convex bornological spaces. The former will be established for X a completely bornological space and will be a consequence of the latter, which will be proved for X a complete convex bornological space. The range space Y has, in both cases, a bornology 'with a net'.

4:4·3 Bornologies with Nets

Let F be a vector space. A *NET* (réseau) in F is a family $\mathcal{R}$ of disks of F:

$$e_{n_1,\dots,n_k} \quad \text{with} \quad k,n_1,n_2,\dots,n_k \in \mathbb{N},$$

satisfying the following condition:

$$(\mathrm{R}):\ F = \bigcup_{n_1=1}^{\infty} e_{n_1} \quad \text{and} \quad e_{n_1,\dots,n_{k-1}} = \bigcup_{n_k=1}^{\infty} e_{n_1,\dots,n_k} \quad \text{for all } k > 1.$$

If $\mathcal{B}$ is a separated convex bornology on F, we say that $\mathcal{R}$ *AND* $\mathcal{B}$ *ARE COMPATIBLE* if the following two properties are verified:

(BR.1): *For every sequence (n_k) of integers, there exists a sequence (ν_k) of positive reals such that, for each $f_k \in e_{n_1,\dots,n_k}$ and for each $\mu_k \in [0,\nu_k]$, the series $\sum_{k=1}^{\infty} \mu_k f_k$ converges bornologically in $(F,\mathcal{B})$ and its sum satisfies $\sum_{k=k_0}^{\infty} \mu_k f_k \in e_{n_1,\dots,n_{k_0}}$ for every $k_0 \in \mathbb{N}$.*

(BR.2): *For every pair $((n_k),(\lambda_k))$ consisting of a sequence (n_k) of positive integers and of a sequence (λ_k) of positive reals, the set $\bigcap_{k=1}^{\infty} \lambda_k e_{n_1,\dots,n_k}$ is bounded in $(F,\mathcal{B})$.*

We say that a *CONVEX BORNOLOGICAL SPACE* $(F,\mathcal{B})$ *HAS A NET*, or that *its BORNOLOGY HAS A NET*, if there exists in F a net $\mathcal{R}$ compatible with $\mathcal{B}$. In this case we also say that $\mathcal{R}$ is a net in $(F,\mathcal{B})$ and that $(F,\mathcal{B})$ is a *SPACE WITH A NET*.

4:4·4 Fundamental Examples of Spaces with Nets

EXAMPLE (1): *If F is a Fréchet space, then ${}^bF$ has a net:* Let (V_n) be a decreasing base of disked neighbourhoods of 0 in F. For every k-tuple $(n_1,\ldots,n_k)$ put:

$$e_{n_1,\ldots,n_k} = n_1V_1 \cap n_kV_k;$$

then the family $(e_{n_1,\ldots,n_k})$ is a net in ${}^bF$. In fact, since every neighbourhood of 0 is absorbent, Condition (R) is trivially satisfied. Let us verify (BR.1). If (n_k) is a given sequence of positive integers, put $\nu_k = (1/2^k n_k)$. Then for every sequence (μ_k), with $\mu_k \in [0,\nu_k]$, and for every $f_k \in e_{n_1,\ldots,n_k}$, the series $\sum_k \mu_k f_k$ satisfies Cauchy's criterion in F, whence it converges topologically in F and hence bornologically in ${}^bF$, since F is metrizable (Proposition (3) of Section 1:4). Moreover:

$$\sum_{k=k_0}^{\infty} \mu_k f_k \in \left(\sum_{k=k_0}^{\infty} 2^{-k}\right) e_{n_1,\ldots,n_{k_0}} \subset e_{n_1,\ldots,n_{k_0}},$$

and so Condition (BR.1) is satisfied. Finally, to show that Condition (BR.2) holds, let (n_k) be a sequence of integers and let (λ_k) be a sequence of positive real numbers. If $A = \bigcap_{k=1}^{\infty} \lambda_k e_{n_1,\ldots,n_k}$, then $A = \bigcap_{k=1}^{\infty} \lambda_k(n_1V_1 \cap \ldots \cap n_kV_k)$, hence A is absorbed by each V_n and, therefore, bounded in ${}^bF$.

EXAMPLE (2): *Let F be a separated convex bornological space, the inductive limit of an increasing sequence (F_n) of convex bornological spaces with nets, the canonical maps $F_n \to F_{n+1}$ being injective. Then F has a net.*

Let $F = \bigcup_{n=1}^{\infty} F_n$ and for every $n \in \mathbb{N}$, let $\mathcal{R}_n$ be a net in F_n:

$$\mathcal{R}_n = \{e^{(n)}_{n_1,\ldots,n_k} : k, n_1, \ldots, n_k \in \mathbb{N}\}.$$

Put:

$$e_{n_1} = F_{n_1}, \qquad n_1 \in \mathbb{N},$$

and:

$$e_{n_1,\ldots,n_k} = e^{(n_1)}_{n_2,\ldots,n_k}, \qquad k > 1, \quad n_1,\ldots,n_k \in \mathbb{N}.$$

It follows immediately from the definitions that the sequence

$(e_{n_1,\dots,n_k})$ is a net in F, since the series in (BR.1) only differ from those considered in the nets $\mathcal{R}_n$ by the addition of one element.

EXAMPLE (3): *Every complete convex bornological space with a countable base has a net:* as follows from Examples (1,2).

4:4·5 The Bornologically Closed Graph Theorem and Its Consequences

THEOREM (1): *Let E and F be convex bornological spaces such that E is complete and F has a net. Every linear map $u:E\to F$ with a bornologically closed graph in $E\times F$ is bounded.*

Before proving this Theorem, we give its most important Corollaries.

COROLLARY (1): *Let E and F be separated locally convex spaces. Suppose that E is completely bornological and that ${}^bF$ has a net. Every linear map $u:E\to F$, whose graph is sequentially closed in $E\times F$, is continuous.*

Proof: Let us recall that a subset A of a separated topological vector space X is sequentially closed if it contains the limit of every sequence in A which converges in X. A closed subset of X is therefore sequentially closed and a sequentially closed subset of X is b-closed in ${}^bX$, since every bornologically convergent sequence is also topologically convergent. Now let us apply these assertions to the space $X = E\times F$ endowed with the product topology. Then, by assumption, the graph of u is b-closed in ${}^b(E\times F) = {}^bE\times {}^bF$. Since E is completely bornological, there exists a complete convex bornological space E_1 such that $E = {}^tE_1$. Thus the identity $E_1\to {}^bE$ is bounded, whence so is the identity $E_1\times {}^bF\to {}^bE\times {}^bF$. The graph of u is then b-closed in $E_1\times {}^bF$ and, by Theorem (1), u is bounded from E_1 to ${}^bF$, hence continuous from E to F.

COROLLARY (2): *Let E and F be Fréchet spaces. Every linear map $u:E\to F$ with closed graph in $E\times F$ is continuous.*

Proof: A Fréchet space X is completely bornological and ${}^bX$ has a net (Example (1)). The Corollary is then an immediate consequence of Corollary (1).

COROLLARY (3): *Let E and F be complete convex bornological spaces and suppose that the bornology of F has a countable base. Every linear map $u:E\to F$, whose graph is b-closed in $E\times F$, is bounded.*

Proof: This follows from Theorem (1) since F has a net (Example (3)).

4:4·6 Proof of Theorem (1)

(a): It suffices to prove the Theorem for E a Banach space. In fact, suppose the Theorem proved in this case and let E be an arbitrary complete convex bornological space. If B is a bounded disk in E, which we may assume to be completant, then E_B is a Banach space and the canonical map $\pi_B : E_B \to E$ is evidently bounded. The restriction of u to $\pi_B(E_B)$ is the map $u \circ \pi_B$, whose graph is b-closed in $E_B \times F$. By hypothesis $u \circ \pi_B$ is bounded and hence $u(B)$ is bounded in F. Since B is arbitrary, the boundedness of u follows.

(b): Hence suppose that E is a Banach space with unit ball B. We shall show that there exists a sequence (n_k) of integers such that $u(B)$ is absorbed in each $e_{n_1,\dots,n_k}$. Granting this for the moment, it follows that there exists a sequence (α_k) of real numbers such that $u(B) \subset \bigcap_{k=1}^{\infty} \alpha_k e_{n_1,\dots,n_k}$ and since the latter set is bounded in F, by (BR.2), we conclude that $u(B)$ is bounded in F.

(c): Existence of the sequence (n_k): By hypothesis $F = \bigcup_{n_1=1}^{\infty} e_{n_1}$ and hence $E = u^{-1}(F) = \bigcup_{n_1=1}^{\infty} u^{-1}(e_{n_1})$. Since E is a Baire space, we can find an integer n_1 for which $u^{-1}(e_{n_1})$ is not meagre in E (*cf.* J. Dieudonné [2], Chapter XII, §16, 12.16.1). Now $e_{n_1} = \bigcup_{n_2=1}^{\infty} e_{n_1,n_2}$, hence $u^{-1}(e_{n_1}) = \bigcup_{n_2=1}^{\infty} u^{-1}(e_{n_1,n_2})$ and again $u^{-1}(e_{n_1,n_2})$ is not meagre in E for some integer n_2; by induction, we can find a sequence (n_k) of integers such that $u^{-1}(e_{n_1,\dots,n_k})$ is not meagre in E. It will suffice to show that every non-meagre set of the form $u^{-1}(e_{n_1,\dots,n_{k_0}})$ absorbs B. Consider the sequence $(n_k : k \geq k_0)$; by (BR.1) there exists a sequence $(\lambda_k : k \geq k_0)$ of positive real numbers such that, whenever $\mu_k \in [0,\lambda_k]$ and $f_k \in e_{n_1,\dots,n_k}$, the series $\sum_{k=k_0}^{\infty} \mu_k f_k$ converges bornologically in F and $\sum_{k=k_0}^{\infty} \mu_k f_k \in e_{n_1,\dots,n_{k_0}}$. Let ε be a given positive number; we can choose the sequence (λ_k) so that $\sum_{k=k_0+1}^{\infty} \lambda_k \leq \varepsilon$. If $A_k = \lambda_k u^{-1}(e_{n_1,\dots,n_k})$, then there is a point a_k in the interior of $\bar{A}_k$ and hence $a_k + \rho_k B \subset \bar{A}_k$ for some number ρ_k. We may assume that $a_k \in A_k$; in fact, since $a_k \in \bar{A}_k$, we can find $a_k' \in A_k$ such that $(a_k' - a_k) \in \frac{1}{2}\rho_k B$. Then:

$$a_k' + \tfrac{1}{2}\rho_k B = (a_k' - a_k) + (a_k + \tfrac{1}{2}\rho_k B) \subset a_k + \rho_k B \subset \bar{A}_k,$$

and we can replace a_k by a_k'. We may also suppose that $\rho_k \leq (1/k)$. For $k = k_0$ we have:

$$\rho_{k_0} B \subset \bar{A}_{k_0} - a_{k_0} \subset 2\bar{A}_{k_0},$$

and the proof will be complete if we show that:

$$\overline{u^{-1}(e_{n_1,\dots,n_{k_0}})} \subset (1 + 2\varepsilon)u^{-1}(e_{n_1,\dots,n_{k_0}}) \tag{1}$$

for then $\rho_{k_0}B \subset 2(1 + 2\varepsilon)\lambda_{k_0}u^{-1}(e_{n_1,\dots,n_{k_0}})$. In order to prove (1) let $x \in \overline{u^{-1}(e_{n_1,\dots,n_{k_0}})}$; then there is an element $y_{k_0} \in u^{-1}(e_{n_1,\dots,n_{k_0}})$ such that $(x - y_{k_0}) \in \rho_{k_0+1}B$ and hence:

$$(x - y_{k_0} + a_{k_0+1}) \in \rho_{k_0+1}B + a_{k_0+1} \subset \bar{A}_{k_0+1} = \lambda_{k_0+1}\overline{u^{-1}(e_{n_1,\dots,n_{k_0+1}})}.$$

Thus we can find $y_{k_0+1} \in \lambda_{k_0+1}u^{-1}(e_{n_1,..,n_{k_0+1}})$ such that $(x - y_{k_0} + a_{k_0+1} - y_{k_0+1}) \in \rho_{k_0+2}B$, that is to say, $(x - (y_{k_0} + y_{k_0+1}) + a_{k_0+1}) \in \rho_{k_0+2}B$. Inductively, for every $k > k_0$ there exists $y_k \in \lambda_k u^{-1}(e_{n_1,\dots,n_k})$ such that, for $N > k_0$:

$$x - \sum_{k=k_0}^{N} y_k + \sum_{k=k_0+1}^{N} a_k \in \rho_{N+1}B. \tag{2}$$

Since $\rho_N \to 0$, the left hand side of (2) converges to 0; we show that its inverse under u converges bornologically in F. Let $z_k = u(y_k)$ and $b_k = u(a_k)$; then:

$$z_{k_0} = u(y_{k_0}) \in e_{n_1,\dots,n_{k_0}}, \qquad b_{k_0} = u(a_{k_0}) \in \lambda_{k_0}e_{n_1,\dots,n_{k_0}},$$

and $z_k, b_k \in \lambda_k e_{n_1,\dots,n_k}$ for $k > k_0$. Now (BR.1) implies that the series $\sum_{k=k_0}^{\infty} z_k$ and $\sum_{k=k_0}^{\infty} b_k$ converge bornologically in F. Moreover, since $e_{n_1,\dots,n_k} \subset e_{n_1,\dots,n_{k_0}}$ for $k > k_0$, we have:

$$\sum_{k=k_0}^{\infty} z_k = z_{k_0} + \sum_{k=k_0+1}^{\infty} z_k \in e_{n_1,\dots,n_{k_0}} + \varepsilon e_{n_1,\dots,n_{k_0}},$$

$$\sum_{k=k_0+1}^{\infty} b_k \in \varepsilon e_{n_1,\dots,n_{k_0}},$$

and hence:

$$y = \sum_{k=k_0}^{\infty} z_k - \sum_{k=k_0+1}^{\infty} b_k \in (1 + 2\varepsilon)e_{n_1,\dots,n_{k_0}}.$$

Thus the image under u of the left hand side of (2) converges to $u(x) - y$. Since the graph of u is b-closed in $E \times F$, we must have:

$$u(x - \sum_{k=k_0}^{\infty} y_k + \sum_{k=k_0+1}^{\infty} a_k) = u(x) - y,$$

hence $u(x) - y = u(0) = 0$ and, consequently, $x \in u^{-1}(y) \in (1 + 2\varepsilon)\times u^{-1}(e_{n_1,\dots,n_{k_0}})$, which proves (1). The proof of Theorem (1) is now complete.

4:4·7 Isomorphism Theorems

THEOREM (2): *Let E and F be convex bornological spaces such that E is complete and F has a net. Every bounded linear bijection $v:F \to E$ is a bornological isomorphism.*

Proof: The map $u = v^{-1}:E \to F$ is a linear map of a complete convex bornological space into a convex bornological space with a net. The bornological isomorphism $(x,y) \to (y,x)$ of $E \times F$ onto $F \times E$ maps the graph of u onto the graph of v. The latter is b-closed in $F \times E$ since v is bounded; hence the graph of u is b-blosed in $E \times F$ and u is bounded by Theorem (1).

COROLLARY (1): *Let E and F be complete convex bornological spaces with a countable base. Every bounded linear bijection of E onto F is a bornological isomorphism.*

In fact, both E and F are complete and have nets.

COROLLARY (2): *Let E and F be Fréchet spaces. Every continuous linear bijection u of E onto F is a topological isomorphism.*

Proof: u is bounded, hence is a bornological isomorphism of ${}^bE$ onto ${}^bF$ (Theorem (2)), because ${}^bE$ and ${}^bF$ are complete and have nets. However, this isomorphism is also topological, since E and F are metrizable and hence bornological.

CHAPTER V

'TOPOLOGY—BORNOLOGY': EXTERNAL DUALITY

I — THE FUNDAMENTAL PRINCIPLES OF DUALITY

Let (F,G) be a 'pair of vector spaces in duality'; to every convex bornology (resp. locally convex topology) on either F or G, 'compatible with this duality', there corresponds by *polarity* a locally convex topology (resp. convex bornology) on the other space. This is the first aspect of the *external duality between topology and bornology* whose general scheme is described in Section 5:1. All separated locally convex topologies on a vector space can be obtained by this general method (Theorem (3) of Section 5:1) which, therefore, presents itself as a universal method Such a result is the most important one in Section 5:1 and will be used constantly thereafter.

The second aspect of the external duality can be expressed as follows: Given a separated locally convex space E, one compares two natural bornologies on its dual E': the equicontinuous bornology and the bornology of equibounded sets. This comparison is carried out in Section 5:2, where we show how it leads to the 'Banach-Steinhaus Theorem' and the notions of barrelled or infrabarrelled spaces, all of which are very important.

In Section 5:3, the completeness of the equicontinuous bornology in a topological dual is established. This is a basic result: it enables us to identify in every dual E' a completely bornological topology directly related to the topology of E (Theorem (1)) and it also implies 'Mackey's Theorem' (Corollary (1) to Theorem (1)).

Section 5:4 establishes the completeness of the natural topology on a bornological dual. It is by appealing to this result that one proves in practice the completeness of the most common dual spaces.

Finally, Section 5:5 investigates the external duality between bounded linear maps and continuous ones via the formation of dual maps, which is one of the fundamental operations in Analysis.

5:0 PRELIMINARIES: THE HAHN-BANACH THEOREM AND ITS CONSEQUENCES

In this Section we collect the necessary preliminaries for the study of the external duality between topology and bornology, i.e. the Hahn-Banach Theorem (in its analytic and geometric forms), the notion of a pair of vector spaces in duality, the definition of a weak topology associated with a duality, the notion of a polar set and the Bipolar Theorem. All these theorems are clearly stated, together with those consequences that will be needed later. However, we shall not give their proofs, for which the reader is referred, for example, to N. Bourbaki [3] or L. Schwartz [1].

5:0·1 The Hahn-Banach Theorem and the Existence of Non-Zero Continuous Linear Functionals

THEOREM (1): *Consider a vector space E over $\mathbb{K}$, a semi-norm p on E and a subspace F of E. If u is a linear functional defined on F and such that $|u(x)| \leqslant p(x)$ for all $x \in F$, then there exists a linear functional $\tilde{u}$ on E such that $\tilde{u}(x) = u(x)$ for all $x \in F$ and $|\tilde{u}(x)| \leqslant p(x)$ for all $x \in E$.*

This Theorem is known as the 'analytic form' of the Hahn-Banach Theorem. As the reader will notice, it holds for every vector space, which *a priori* is endowed with neither a topology nor a bornology. What is essential is the *existence of a semi-norm p on E* satisfying the conditions of the statement. Now a locally convex topology on a vector space E has precisely the advantage of implying the existence of such a semi-norm, thus yielding:

COROLLARY (1): *Let E be a locally convex space and let F be a subspace of E equipped with the induced topology. For every continuous linear functional u on F, there exists a continuous linear functional $\tilde{u}$ on E such that $\tilde{u}(x) = u(x)$ for all $x \in F$.*

Proof: Let D be the unit ball of $\mathbb{K}$; the linear functional $u: F \to \mathbb{K}$ being continuous, $u^{-1}(D)$ is a neighbourhood of 0 in F. Since F has the topology induced by E and E is locally convex, there exists a disked neighbourhood V of 0 in E such that $u^{-1}(D) \supset V \cap F$. The gauge p of V is a semi-norm on E, since V is an absorbent disk in E. If $x \in V \cap F$, then $u(x) \in D$ and hence $|u(x)| \leqslant 1$. Let y be an arbitrary element of F. For every $\varepsilon > 0$, $y/(p(y) + \varepsilon) \in V \cap F$, hence $|u(y/(p(y) + \varepsilon))| \leqslant 1$, i.e. $|u(y)| \leqslant p(y) + \varepsilon$. Since ε is arbitrary, $|u(y)| \leqslant p(y)$ for all $y \in F$. Thus the conditions of Theorem (1) are satisfied and we deduce the existence of a linear functional $\tilde{u}$ on E, extending u and such that $|\tilde{u}(x)| \leqslant p(x)$ for all $x \in E$, and this inequality means precisely that $\tilde{u}$ is continuous on E.

COROLLARY (2): *Let E be a separated locally convex space and let $x \in E$, $x \neq 0$. There exists a continuous linear functional u on E such that $u(x) \neq 0$.*

Proof: Since E is separated, the subspace $\{0\}$ is closed and hence there exists a disked neighbourhood V of 0 with $x \notin V$. If p is the gauge of V, then p is a semi-norm on E and $p(x) \neq 0$. Let $F = \mathbb{K}x$ be the subspace spanned by x and define a linear functional v on F by $v(\lambda x) = \lambda$. We have:

$$|v(\lambda x)| = |\lambda| = \frac{p(\lambda x)}{p(x)},$$

i.e. $|v(y)| \leqslant q(y)$ for all $y \in F$, where $q(y) = p(y)/p(x)$ is a semi-norm on E. By virtue of Theorem (1), there exists a linear functional u on E such that $u(\lambda x) = v(\lambda x)$ for all $\lambda \in \mathbb{K}$ (hence $u(x) = v(x) = 1 \neq 0$) and, moreover, $|u(y)| \leqslant q(y)$ for all $y \in E$, which ensures the continuity of u.

5:0·2 The Hahn-Banach Theorem and the Closure of a Convex Set

Let us recall that *a HYPERPLANE in a vector space E is the kernel of a linear functional on E.* Then Theorem (1) can be stated in the following equivalent form, called the 'geometric form' of the Hahn-Banach Theorem.

THEOREM (2): *Let E be a topological vector space, let A be a non-empty convex open subset of E and let F be a subspace of E not intersecting A. There exists a closed hyperplane in E, containing F and not intersecting A.*

From this Theorem we shall deduce three consequences which, together with Corollary (2) to Theorem (1), are the only statements that will be used.

COROLLARY (1): *In a locally convex space, every closed subspace is the intersection of all closed hyperplanes containing it.*

Proof: In fact, let F be a closed subspace of a locally convex space E. If $x \notin F$, there exists a convex open neighbourhood V of x whose intersection with F is empty (since the interior of a convex set is convex). Then Theorem (2) ensures the existence of a closed hyperplane in E containing F and having empty intersection with V. *A fortiori*, such a hyperplane does not contain x and the assertion follows.

In order to state Corollary (2) we give the following Definition. Let E be a real topological vector space; a *CLOSED HALF-SPACE* in E is a subset of the form $\{x \in E; f(x) \leqslant \alpha\}$ or $\{x \in E; f(x) \geqslant \alpha\}$, with f a continuous linear functional on E and α a real number. An important consequence of Theorem (2) is the following, which we state without proof:

COROLLARY (2): *Let E be a locally convex space over $\mathbb{R}$. The closure of a convex set $A \subset E$ is the intersection of the closed half-spaces containing A.*

COROLLARY (3): *Let E be a separated locally convex space and let F be a subspace of E. Then F is dense in E if and only if every continuous linear functional on E vanishing on F is identically zero on E.*

Proof: The necessity is obvious, the kernel of a continuous linear functional being closed. To prove the sufficiency, let $\bar{F}$ be the closure of F in E; if $\bar{F} \neq E$, then there exists $x \in E$ with $x \notin \bar{F}$ and so, by Corollary (1), there is a closed hyperplane H containing $\bar{F}$ and such that $x \notin H$. Since H is the kernel of a continuous linear functional u on E, u vanishes on $\bar{F}$ but not at x, whence not on E, which is contrary to the assumption.

REMARK (1): Let E be a separated locally convex space with topology $\mathcal{T}_0$ and let E' be its dual. If $\mathcal{T}_1$ is another locally convex topology on E such that the dual of $(E,\mathcal{T}_1)$ is again E', then the closure of a convex subset of E, in particular, of a subspace, is the same for both $\mathcal{T}_0$ and $\mathcal{T}_1$. Indeed, by Corollary (2), the closure of a convex set depends only on E'. All topologies on E yielding E' as a dual will be characterised in Chapter VI by the 'Mackey-Arens Theorem'.

5:0˙3 Dual Pairs

5:0˙3(a)

DEFINITION: *Let F and G be vector spaces over* $\mathbb{K}$. *If a bilinear form B is defined on $F \times G$, $(x,y) \to B(x,y)$, we say that F and G are in* DUALITY VIA THE BILINEAR FORM B, *or that (F,G) is a* DUALITY WITH BILINEAR FORM B. *The duality between F and G is called a* DUALITY SEPARATED IN F *if for every $x \in F$, $x \neq 0$, there exists $y \in G$ such that $B(x,y) \neq 0$. Similarly, the duality is a* DUALITY SEPARATED IN G *if for every $y \in G$, $y \neq 0$, there exists $x \in F$ such that $B(x,y) \neq 0$. The duality (F,G) will simply be called a* SEPARATED DUALITY *if it is separated in both F and G.*

5:0˙3(b)

EXAMPLE (1): Let E be a vector space and let E^* be its *ALGEBRAIC DUAL*, i.e. the vector space of all linear functionals on E. For every $x^* \in E^*$ and $x \in E$ we denote by $\langle x,x^*\rangle$ the scalar $x^*(x)$, i.e. the value of the linear functional x^* at the point x. The map $E \times E^* \to \mathbb{K}$ defined by:

$$(x,x^*) \to \langle x,x^*\rangle$$

is a bilinear form on $E \times E^*$ called the *CANONICAL BILINEAR FORM*. This bilinear form induces a separated duality between E and E^*. It is obvious that the duality is separated in E^*. Conversely, if $x \in E$ and $x \neq 0$, let $(e_i)_{i \in I}$ be a Hamel basis in E; then $x = \sum_{i \in I} \lambda_i e_i$, where at least one of the scalars λ_i, e.g. λ_j, is different from 0. Hence, if x^* is the linear functional on E mapping every $y = \sum_{i \in I} \alpha_i e_i$ to the scalar α_j, then $\langle x,x^*\rangle = \lambda_j \neq 0$.

EXAMPLE (2): *Topological Duality*: Let E be a locally convex space and let E' be its *TOPOLOGICAL DUAL*, i.e. the vector space of all continuous linear functionals on E. Since E' is a subspace of E^*, the restriction of the canonical bilinear form induces a duality between E and E'. If E is separated, then Corollary (2) to Theorem (1) ensures that this duality is separated in E, whence it is a separated duality, since it is always separated in E'. Conversely, if such a duality is separated in E, then E is necessarily separated.

EXAMPLE (3): *Bornological Duality*: Let E be a convex bornological space. The set of all bounded linear functionals on E is a vector space called the *BORNOLOGICAL DUAL* of E and *denoted by* $E^\times$. We can induce a duality between E and $E^\times$ by using the canonical bilinear form:

$$(x, x^\times) \to \langle x, x^\times \rangle = x^\times(x),$$

for $x \in E$ and $x^\times \in E^\times$. This duality is called the *BORNOLOGICAL DUALITY between E and $E^\times$*. Since, algebraically, $E^\times = ({}^tE)'$ (the topological dual of E'), we see that the bornological duality between E and $E^\times$ is identical to the topological duality between ${}^tE$ and $({}^tE)'$. Thus, in view of Example (2), this duality, which is always separated in $E^\times$, will be separated in E if (and only if) the topology of ${}^tE$ is separated, a condition which is not always satisfied (*cf*. Exercise 3·E.5). Thus we are led to introduce a new class of convex bornological spaces, called *REGULAR* (or *t-SEPARATED*) *CONVEX BORNOLOGICAL SPACES*: these are exactly those spaces E for which ${}^tE$ is separated or, equivalently, such that $E^\times$ separates E. The following regularity criterion is obvious.

PROPOSITION (1): *A convex bornological space E is regular if and only if there is on E a separated locally convex topology compatible with the bornology of E.*

In fact, such a topology is coarser than ${}^tE$.

5:0·4 Weak Topologies Defined by a Duality

Let F and G be vector spaces in duality via a bilinear form B, which will be denoted by:

$$B(x, y) = \langle x, y \rangle .$$

For every $y \in G$, the map $x \to |\langle x, y\rangle|$ is a semi-norm on F, denoted by p_Y. The locally convex topology defined on F by the family $\{p_Y; y \in G\}$ of semi-norms is called the *WEAK TOPOLOGY ON F DEFINED BY THE DUALITY* (F,G) and is *denoted by* $\sigma(F,G)$. The form of neighbourhoods of 0 for this topology will be given later on, in the context of a general and universal method for constructing locally convex topologies. Similarly, we can define the *WEAK TOPOLOGY* $\sigma(G,F)$ *ON* G by symmetry.

Note that the topology $\sigma(F,G)$ is separated if and only if the duality (F,G) is separated in F.

PROPOSITION (2): *Let (F,G) be a separated duality with bilinear form $(x,y) \to \langle x,y\rangle$. We give F the topology $\sigma(F,G)$. Then for every $y \in G$, the map $x \to \langle x,y\rangle$ is a continuous linear functional on F and, conversely, for every continuous linear functional u on F there exists a unique $y \in G$ such that $u(x) = \langle x,y\rangle$ for all $x \in F$.*

Thus, in view of the uniqueness of the element y corresponding to the continuous linear functional u, we may identify G with the dual of the space F endowed with $\sigma(F,G)$.

Proposition (2) has, of course, a symmetric analogue for $(G,\sigma(G,F))$. In particular, if E is a locally convex space we may consider the topology $\sigma(E,E')$ on E and the topology $\sigma(E',E)$ on E'. The topology $\sigma(E,E')$ is, clearly, always coarser than the given topology on E and is called the WEAK TOPOLOGY *of* E. Proposition (2) then asserts that the space E, when endowed with its weak topology, is always 'reflexive' in a sense that will be made precise in the following Chapter.

5:0·5 Polarity

Let (F,G) be a duality with bilinear form $(x,y) \to \langle x,y\rangle$. For every non-empty subset A of F we define the POLAR $A°$ *of* A *(relative to* (F,G)*)* as the set of all elements $y \in G$ such that $|\langle x,y\rangle| \leq 1$ for all $x \in A$. The polar of a subset of G is defined similarly.

Polar sets have the following elementary Properties:

(i): $A \subset B \Rightarrow A° \supset B°$.

(ii): $(A \cup B)° = A° \cap B°$.

(iii): *For every* $\lambda \in \mathbb{K}$, $\lambda \neq 0$, $(\lambda A)° = (1/\lambda)A°$.

(iv): $A°$ *is always disked and closed for* $\sigma(G,F)$.

(v): $A° = (\Gamma(A))°$.

(vi): *If* G_1 *is a subspace of* G*, then the restriction to* $F \times G_1$ *of the bilinear form of* (F,G) *induces a duality between* F *and* G_1 *and we have, for every* $A \subset F$:

$$A^{\circ}_{G_1} = A^{\circ}_{G} \cap G_1,$$

where $A^{\circ}_{G_1}$ *(resp.* A°_{G}*) is the polar of* A *in* G_1 *(resp.* G*).*

(vii): *For every set* $A \subset F$*, the polar* $(A°)°$ *of* $A°$ *in* F *is called the* BIPOLAR *of* A *and is denoted by* $A^{\circ\circ}$*. Clearly* $A \subset A^{\circ\circ}$.

It is very important to know the conditions under which we have equality, and these are given by the following Theorem.

THEOREM (3): (Bipolar Theorem): *Let* (F,G) *be a duality. If* A *is a non-empty subset of* F*, then* $A^{\circ\circ}$ *is the closure for* $\sigma(F,G)$ *of the disked hull of* A.

From this Theorem we deduce:

COROLLARY (1): *Let (F,G) be a duality and let A be a non-empty subset of F. Then $A = A^{\circ\circ}$ if and only if A is a disk which is closed for $\sigma(F,G)$.*

In particular, let E be a locally convex space with dual E'. The given topology on E and the weak topology $\sigma(E,E')$ yield the same dual (Proposition (2)). Thus these topologies have the same closed half-spaces, hence the same closed convex sets (Corollary (2) to Theorem (2)) and we have:

COROLLARY (2): *Let E be a locally convex space and let A be a non-empty subset of E. Then $A = A^{\circ\circ}$ if and only if A is a closed disk.*

5:1 THE EXTERNAL DUALITY BETWEEN TOPOLOGY AND BORNOLOGY

5:1˙1 The Polar Topology of a Bornology

5:1˙1(a)

THEOREM (1): *Let (F,G) be a separated duality and let $\mathcal{B}$ be a bornology on G compatible with the topology $\sigma(G,F)$. Denote by:*

$$\mathcal{B}^{\circ} = \{B^{\circ}; B \in \mathcal{B}\},$$

the family of polars in F of elements of $\mathcal{B}$ with respect to the duality (F,G). Then $\mathcal{B}^{\circ}$ is a base for a separated locally convex topology on F.

Proof: $\mathcal{B}^{\circ}$ is a filter base, since $0 \in A^{\circ}$, $A^{\circ} \cap B^{\circ} = (A \cup B)^{\circ}$ whenever A and B belong to $\mathcal{B}$, and $\mathcal{B}$ is directed under inclusion. Clearly $\mathcal{B}$ consists of disks (*cf.* Subsection 5:0˙5) and is stable under homothetic transformations. Hence it suffices to show that the sets B° are absorbent. Let, then, $A \in \mathcal{B}$ and $u \in F$; since $\mathcal{B}$ is compatible with $\sigma(G,F)$, $u(A)$ is bounded in $\mathbb{K}$ and we can find a $\lambda > 0$ such that $|u(x)| \leqslant \lambda$ for all $x \in A$. This implies that $(u/\lambda) \in A^{\circ}$, hence that A° absorbs u. Thus $\mathcal{B}^{\circ}$ defines a locally convex topology on F which is separated, since $\mathcal{B}$ covers G and hence $\bigcap_{B \in \mathcal{B}} B^{\circ} = \{0\}$.

DEFINITION (1): *With the notation of Theorem (1), the topology defined on F by $\mathcal{B}^{\circ}$ is called the $\mathcal{B}$-TOPOLOGY on F or the POLAR TOPOLOGY on F OF THE BORNOLOGY $\mathcal{B}$.*

REMARK (1): *For every set $A \subset G$, the polar of A in F is the same as the polar of the closure for $\sigma(G,F)$ of the disked hull $\Gamma(A)$ of A.* In fact, since A and $\Gamma(A)$ have the same polar (Section 5:0 it suffices to show that if B is a subset of G, then the polars of B and $\bar{B}$ (the closure of B for $\sigma(G,F)$) are the same. Now $B^{\circ\circ}$ is closed for $\sigma(G,F)$ and contains B, hence $B \subset \bar{B} \subset B^{\circ\circ}$. By Properties (i,viii) of polar sets in Subsection 5:0˙5, $B^{\circ} \supset (\bar{B})^{\circ} \supset B^{\circ\circ\circ} = B^{\circ}$, and, therefore, $B^{\circ} = (\bar{B})^{\circ}$.

Thus we may assume in Theorem (1) that the elements of $\mathcal{B}$ are closed for $\sigma(G,F)$.

5:1˙1(b) *The Semi-Norms of a $\mathcal{B}$-Topology*

With the notation of Theorem (1), the semi-norms defining the $\mathcal{B}$-topology on F are given by the gauges of the sets B° for $B \in \mathcal{B}$. Since $y \in \lambda A^\circ$ is equivalent to $\sup_{x \in A} |\langle x,y \rangle| \leq \lambda$, we see that the gauge of A° is given by the expression:

$$p_A(y) = \sup_{x \in A} |\langle x,y \rangle|,$$

where $\langle\, ,\rangle$ denotes the bilinear form of the duality (F,G).

A net (in the sense of Subsection 0·C.4˙3) of elements of F which converges to 0 for the $\mathcal{B}$-topology of F converges to 0 uniformly on each member A of $\mathcal{B}$; for this reason, the $\mathcal{B}$-topology is also called the *TOPOLOGY OF UNIFORM CONVERGENCE ON THE BORNOLOGY* $\mathcal{B}$ or *on the members of* $\mathcal{B}$.

5:1˙1(c) *Examples of $\mathcal{B}$-Topologies*

EXAMPLE (1): *Weak Topologies and Finite-Dimensional Bornologies:* Let (F,G) be a separated duality and let $\mathcal{B}$ be the finite-dimensional bornology on G (Subsection 2:9˙4). $\mathcal{B}$ is obviously compatible with every locally convex topology on G, since every neighbourhood of 0 is absorbent. Thus the family $\mathcal{B}^\circ$ of polars in F of members of $\mathcal{B}$ defines a separated locally convex topology on F. This topology is precisely $\sigma(F,G)$. In fact, the semi-norms defining the $\mathcal{B}$-topology are given by expressions of the type:

$$p_A(y) = \sup_{1 \leq i \leq n} |\langle x_i,y \rangle|,$$

where $A = \{x_1,\ldots,x_n\}$ is a finite subset of G. Since, by definition, the semi-norms of $\sigma(F,G)$ are given by:

$$p_x(y) = |\langle x,y \rangle|, \qquad x \in G,$$

we have:

$$p_A(y) = \sup_{1 \leq i \leq n} p_{x_i}(y), \qquad \text{for } A = \{x_1,\ldots,x_n\}.$$

Hence the $\mathcal{B}$-topology is indeed the topology $\sigma(F,G)$ on F.

EXAMPLE (2): *The Natural Topology on a Bornological Dual:* Let E be a regular convex bornological space with bornology $\mathcal{B}$ and let $E^\times$ be the bornological dual of E. Since E and $E^\times$ form a separated duality, we may consider on $E^\times$ the $\mathcal{B}$-topology associated with such a duality. This topology is called the *NATURAL TOPOLOGY OF* $E^\times$.

EXAMPLE (3): *The Strong Topology on a Topological Dual:* Let E be a separated locally convex space, let E' be its topological dual and let $\mathcal{B}$ be the von Neumann bornology of E. $\mathcal{B}$ is compatible with the given topology on E, *a fortiori*, with the weak topology $\sigma(E,E')$. The $\mathcal{B}$-topology on E' is called the *STRONG TOPOLOGY* and *denoted by* $\beta(E',E)$. The space E', endowed with its strong topology, is called the *STRONG DUAL of* E and *denoted by* E'_β. When working in E', the term '*strongly*' will always mean 'relative to the strong topology' and we shall speak of *strongly convergent sequences*, *strongly bounded sets*, etc..

EXAMPLE (4): *The Topology of Compact or Precompact Convergence:* Let E be a separated locally convex space. The *TOPOLOGY OF COMPACT CONVERGENCE* on E' is the $\mathcal{B}$-topology when $\mathcal{B}$ is the compact bornology of E. Similarly, if $\mathcal{B}$ is the precompact bornology of E, we obtain the *TOPOLOGY OF PRECOMPACT CONVERGENCE on* E'

5:1·2 The Polar Bornology of a Topology

DEFINITION (2): *Let* (F,G) *be a separated duality. A separated locally convex topology* $\mathcal{T}$ *on* F *is said to be a* TOPOLOGY COMPATIBLE WITH THE DUALITY (F,G) *if* G *is the topological dual of* $(F,\mathcal{T})$.

This Definition means means two things: first, every element $y \in G$ defines a continuous linear functional on $(F,\mathcal{T})$ by means of the map $x \to \langle x,y \rangle$ for $x \in F$ and, second, every continuous linear functional u on $(F,\mathcal{T})$ is uniquely determined by an element $y \in G$ via the relationship $u(x) = \langle x,y \rangle$ for all $x \in F$.

Thus to say that $\mathcal{T}$ is compatible with the duality (F,G) is equivalent to saying that $\mathcal{T}$ and $\sigma(F,G)$ have the same dual. The topology $\sigma(F,G)$ is, therefore, an example of a separated locally convex topology on F compatible with the duality (F,G). In the next Chapter we shall give a complete characterisation of all topologies compatible with a given duality.

THEOREM (2): *Let* (F,G) *be a duality and let* $\mathcal{T}$ *be a separated locally convex topology on* F *compatible with* (F,G). *Let* $\mathcal{V}$ *be a base of neighbourhoods of* 0 *in* $(F,\mathcal{T})$ *and let:*

$$\mathcal{V}^\circ = \{V^\circ ; V \in \mathcal{V}\}$$

be the family of polars in G *of members of* $\mathcal{V}$. *Then* $\mathcal{V}^\circ$ *is a base for a separated convex bornology on* G *and this bornology is exactly the equicontinuous bornology of* $(F,\mathcal{T})'$.

Proof: Denote by $(F,\mathcal{T})'$ the topological dual of $(F,\mathcal{T})$. By hypothesis $G = (F,\mathcal{T})'$ algebraically and hence it suffices to show that $\mathcal{V}^\circ$ is a base for the equicontinuous bornology on $(F,\mathcal{T})'$. Now every set $H = V^\circ$ is equicontinuous; in fact, if D is the unit ball of $\mathbb{K}$, then for every neighbourhood λD of 0 in $\mathbb{K}$ we have $H^{-1}(\lambda D) = \lambda H^{-1}(D) \supset \lambda V$, which shows that $H^{-1}(\lambda D)$ is a neighbourhood of 0 in $(F,\mathcal{T})$. Conversely, let H be an equicontinuous subset of $(F,\mathcal{T})'$; we show that there exists $V \in \mathcal{V}$ such that $H \subset V^\circ$.

Since $H^{-1}(D)$ is a neighbourhood of 0 in $(F,\mathfrak{T})$, there exists $V \in \mathfrak{V}$ such that $H^{-1}(D) \supset V$ and hence $V^\circ \supset (H^{-1}(D))^\circ$. By definition, $(H^{-1}(D))^\circ \supset H$, hence $V^\circ \supset H$ and the proof of the Theorem is complete.

COROLLARY: *Let E be a separated locally convex space with dual E', let $\mathfrak{V}$ be a base of neighbourhoods of 0 in E and let:*

$$\mathfrak{V}^\circ = \{V^\circ ; V \in \mathfrak{V}\}$$

be the family of polars in E' of members of $\mathfrak{V}$. Then $\mathfrak{V}^\circ$ is a base for the equicontinuous bornology of E'.

Proof: This follows from Theorem (2), since the topology of E is consistent with the duality (E,E').

5:1·3 Original Topology and Polar Topology of the Equicontinuous Bornology

Let E be a separated locally convex space with dual E', let $\mathfrak{K}$ be the equicontinuous bornology of E' and let $\mathfrak{K}^\circ = \{H^\circ ; H \in \mathfrak{K}\}$ be the family of polars in E of members of $\mathfrak{K}$. Consider a base $\mathfrak{V}$ of neighbourhoods of 0 in E consisting of closed disks. For every $V \in \mathfrak{V}$, $V^{\circ\circ} = V$ (Bipolar Theorem: Corollary (2) to Theorem (3) of Section 5:0). But $\mathfrak{V}^\circ = \{V^\circ ; V \in \mathfrak{V}\} = \mathfrak{K}$ (Corollary to Theorem (2)), hence $\mathfrak{K}^\circ = \mathfrak{V}^{\circ\circ} = \mathfrak{V}$. Consequently, $\mathfrak{K}^\circ$ is a base of neighbourhoods of 0 for the original topology on E.

This proves, at the same time, that the bornology $\mathfrak{K}$ is compatible with the weak topology $\sigma(E',E)$ since for every $H = V^\circ$, with $V \in \mathfrak{V}$, the fact that V is absorbent implies that V° is absorbed by the polars of finite subsets of E, i.e. that V° is bounded for $\sigma(E',E)$. We can now state the following Theorem:

THEOREM (3): *Let E be a separated locally convex space with dual E' and let $\mathfrak{K}$ be the equicontinuous bornology of E'. Then the given topology on E is the $\mathfrak{K}$-topology.*

This Theorem asserts that the method of construction of locally convex topologies by taking polars of bornologies is, in a certain sense, universal: every locally convex topology can be obtained in this way. This idea is extremely convenient, since, in practice, it is often very much easier to construct bornologies, than topologies satisfying given conditions.

5:2 Duality between Equicontinuous and Equibounded Sets in a Dual Space

5:2·1

Let E be a separated locally convex space, let $\mathfrak{B}$ be a bornology on E compatible with the topology of E and let E' be the dual of E equipped with the $\mathfrak{B}$-topology. In this Section we investigate the relationship between the equicontinuous bornology of E' and the von Neumann bornology of E' when this space is given the $\mathcal{B}$-topology. We begin with a characterisation of the latter bornology.

PROPOSITION (1): *The von Neumann bornology of the $\mathcal{B}$-topology is the $\mathcal{B}$-bornology.* (*Example* (8) *of Section* 1:3).

Proof: We have to show that a set $H \subset E'$ is bounded for the $\mathcal{B}$-topology if and only if, for every $A \in \mathcal{B}$, $H(A)$ is bounded in $\mathbb{K}$. The condition is *necessary*: Let H be bounded for the $\mathcal{B}$-topology of E'. For every $A \in \mathcal{B}$, A° is a neighbourhood of 0 for the $\mathcal{B}$-topology and hence absorbs H. Thus there is a $\lambda > 0$ with $\lambda A^\circ \supset H$. Then for all $u \in H$ and $x \in A$, $|(1/\lambda)u(x)| \leqslant 1$, i.e. $|u(x)| \leqslant \lambda$ and $H(A)$ is bounded in $\mathbb{K}$. The condition is *sufficient*: Let H be equibounded on each $B \in \mathcal{B}$ and let V be a neighbourhood of 0 in E' for the $\mathcal{B}$-topology. We may assume that $V = A^\circ$, with $A \in \mathcal{B}$. Since $H(A)$ is bounded in $\mathbb{K}$, we have, for some constant $\alpha > 0$:

$$|u(x)| \leqslant \alpha \quad \text{for all} \quad u \in H,\ x \in A,$$

i.e. $|u(x/\alpha)| \leqslant 1$ for $x \in A$, $u \in H$. This implies that $H \subset \alpha A^\circ$, that is, H is absorbed by $V = A^\circ$ and is, therefore, bounded for the $\mathcal{B}$-topology.

PROPOSITION (2): *Each equicontinuous subset of E' is bounded for every $\mathcal{B}$-topology on E', if $\mathcal{B}$ is a bornology on E compatible with the given topology of E.*

Proof: Consider an equicontinuous set $H \subset E'$ and suppose that $H = V^\circ$, with V a neighbourhood of 0 in E (Corollary to Theorem (2) of Section 5:1). A neighbourhood of 0 for the $\mathcal{B}$-topology is of the form A°, with $A \in \mathcal{B}$. Since A is bounded in E, by virtue of the compatibility of $\mathcal{B}$ with the topology of E, V absorbs A and hence $H = V^\circ$ is absorbed by A°, which shows H to be bounded for the $\mathcal{B}$-topology.

Propositions (1,2) together yield:

PROPOSITION (3): *For every equicontinuous subset H of E' and for every bounded subset A of E, $H(A)$ is bounded in* $\mathbb{K}$.

5:2·2 Barrelledness

We have seen in Proposition (2) above that an equicontinuous subset of E' is bounded for every $\mathcal{B}$-topology, or, in other words, that the equicontinuous bornology of E' is finer than any $\mathcal{B}$-bornology on E', $\mathcal{B}$ being a bornology on E compatible with the topology of E. A natural and important problem is to know when the equicontinuous bornology coincides with the $\mathcal{B}$-bornology when $\mathcal{B}$ is one of the following bornologies on E: the finite-dimensional bornology or the von Neumann bornology. The corresponding $\mathcal{B}$-topologies on E' are then the weak topology $\sigma(E',E)$ and the strong topology $\beta(E',E)$, whilst the $\mathcal{B}$-bornologies are their respective von Neumann bornologies (Proposition (1)).

The discussion of this problem will lead us, in a natural way, to the notion of a *barrelled space* on the one hand and to that of an *infra-barrelled space* on the other. Since the arguments are essentially the same in both cases, we shall only deal with the case when $\mathcal{B}$ is the finite-dimensional bornology, the other case being considered in the Exercises.

PROPOSITION (4): *Let E be a separated locally convex space with dual E'. The following assertions are equivalent:*

(i): *Every subset of E', bounded for $\sigma(E',E)$, is equicontinuous;*

(ii): *Every closed absorbent disk in E is a neighbourhood of* 0.

Proof: (i) => (ii): Let D be a closed absorbent disk in E and let D° be its polar in E'. Since D absorbs the finite subsets of E, D° is absorbed by the polars of such sets, i.e. by the neighbourhoods of 0 for $\sigma(E',E)$. Thus D° is bounded for $\sigma(E',E)$, hence equicontinuous by (i) and $D^{\circ\circ} = D$ (Bipolar Theorem) is a neighbourhood of 0 in E.

(ii) => (i): Let H be a subset of E' which is bounded for $\sigma(E',E)$. H is absorbed by the polars of finite subsets of E, hence H° absorbs the finite subsets of E, i.e. it is absorbent in E. But H° is a disk which is closed for $\sigma(E,E')$, hence closed for the original topology of E and, by (ii), is a neighbourhood of 0 in E. It follows that $H^{\circ\circ}$ is an equicontinuous subset of E', hence, *a fortiori*, so is H.

DEFINITION (1): *A locally convex space is called* **BARRELLED** *if it satisfies either of the (equivalent) conditions of Proposition* (4).

Barrelled spaces are characterised by an important theorem called the 'Banach-Steinhaus Theorem'. In order to state it, we need the following Definition. Let E and F be locally convex spaces and let $L(E,F)$ be the space of continuous linear maps of E into F. A subset H of $L(E,F)$ is said to be ***SIMPLY BOUNDED*** if for every $x \in E$, the set $H(x) = \bigcup_{u \in H} u(x)$ is bounded in $\mathbb{K}$. Thus simply bounded sets are the bounded sets for the $\mathcal{B}$-bornology on $L(E,F)$, where $\mathcal{B}$ is the finite-dimensional bornology on E. It is clear that every equicontinuous subset of $L(E,F)$ is simply bounded; the converse is true if E is barrelled.

THEOREM (1): (Banach-Steinhaus Theorem): *If E is barrelled, every simply bounded subset of $L(E,F)$ is equicontinuous.*

Proof: Let H be a simply bounded subset of $L(E,F)$ and let V be a closed disked neighbourhood of 0 in F. The set $H^{-1}(V) = \bigcap_{u \in H} u^{-1}(V)$ is a closed disk in E which is absorbent, since H is simply bounded. But E is barrelled, hence $H^{-1}(V)$ is a neighbourhood of 0, and, therefore, H is equicontinuous.

COROLLARY: *Let E be a barrelled space and let (u_n) be a sequence of continuous linear maps of E into a locally convex space F. Suppose that for every $x \in E$, the sequence $(u_n(x))$ converges to an element $u(x)$ in F and let $u: E \to F$ be the map thus defined. Then u is a continuous linear map.*

Proof: Since for every $x \in E$, $(u_n(x))$ is a convergent sequence in F, the sequence (u_n) is simply bounded in $L(E,F)$, hence equicontinuous by Theorem (1). The map u is obviously linear. Let V be a closed neighbourhood of 0 in F; since (u_n) is equicontinuous, $U = \bigcup_{n=1}^{\infty} u_n^{-1}(V)$ is a neighbourhood of 0 in E, whence $u_n(U) \subset V$ and the continuity of u is ensured.

The most important example of a barrelled space is given by the following Proposition.

PROPOSITION (5): *Every completely bornological space is barrelled.*

Proof: Let D be a closed absorbent disk in E; we have to show that D is a neighbourhood of 0 in E. The space E, being completely bornological, is of the form $E = {}^tE_1$, with E_1 a complete convex bornological space; hence it suffices to show that D absorbs every completant bounded disk B of E_1. This will be a consequence of the following general Lemma.

LEMMA (1): *Let E be a complete convex bornological space. Every bornologically closed and absorbent disk of E_1 is bornivorous.*

Proof: Let K_1 be a disk in E_1 as in the statement of the Lemma and let B be a completant bounded disk in E_1. Denote by F the Banach space $(E_1)_B$ and put $K = K_1 \cap F$; it is enough to show that K absorbs B. Now K is a closed absorbent disk in F. Since $F = \bigcup_{n=1}^{\infty} nK$ and F is a Banach, hence Baire, space, one of the sets nK must have an interior point in F, whence K itself must have an interior point x_0. Then there exists a neighbourhood V of 0 in E_B such that $(x_0 + V) \subset K$, which implies that $V \subset K + K = 2K$ and hence that K is a neighbourhood of 0 in F. Thus K absorbs B and the Lemma is proved.

5:2·3

In the same order of ideas of this Section, we have the following very useful result.

PROPOSITION (6): *Let E be a bornological locally convex space and let E_1 be an arbitrary convex bornological space such that $E = {}^tE_1$. A subset H of E' is equicontinuous if and only if H is equibounded on each bounded subset of E_1.*

Proof: Let $\mathcal{B}$ be the convex bornology of E_1; by virtue of Proposition (3), every equicontinuous subset of E' is equibounded on each $A \in \mathcal{B}$. For the converse, let H be a subset of E' which is equibounded on each $A \in \mathcal{B}$. Then H is bounded in E' for the $\mathcal{B}$-topology (Proposition (1)) and hence is absorbed by the polars of members of $\mathcal{B}$; it follows that H° absorbs all members of $\mathcal{B}$. Now H°, being a disk in E, is a neighbourhood of 0 in E, hence $H^{\circ\circ}$, and *a fortiori* H, is equicontinuous in E'.

COROLLARY: *Let E be a bornological locally convex space. Every strongly bounded subset of E' is equicontinuous.*

Proof: A subset of E' is strongly bounded if and only if it is equibounded on each bounded subset of E (Proposition (1)). Thus if $E = {}^tE_1$, with E_1 a convex bornological space, then a strongly bounded subset of E' is, *a fortiori*, equibounded on each bounded subset of E_1 and the assertion follows from Proposition (6).

5:2·4

Another important consequence of Lemma (1) above concerns the relationship between 'weakly bounded' and 'strongly bounded' sets (see also Exercise 5·E.2).

PROPOSITION (7): *Let E be a separated, bornologically complete, locally convex space. Every weakly bounded subset E' is strongly bounded.*

Proof: If H is a weakly bounded subset of E', its polar H° in E is a disk which is closed for $\sigma(E,E')$, hence closed for the topology of E. Moreover, H° is absorbent because H is weakly bounded. Since ${}^bE$ is complete, Lemma (1) ensures that H° is bornivorous in E. It follows that $H^{\circ\circ}$ is absorbed by the polars of bounded subsets of E, in other words, $H^{\circ\circ}$ is strongly bounded and, *a fortiori*, so is H.

5:3 COMPLETENESS OF THE EQUICONTINUOUS BORNOLOGY: COMPLETELY BORNOLOGICAL TOPOLOGY ON A DUAL SPACE

In this Section we prove the completeness of the equicontinuous bornology on the dual of a separated locally convex space. This result will be strengthened in the next Chapter by a property of 'weak compactness', but it can easily be proved here. The completeness of the equicontinuous bornology will imply the existence of a natural completely bornological topology on the topological dual of a separated locally convex space, the interest of such a topology having been made precise in Section 4:3. As a consequence, we shall deduce the identity of all bornologies on E associated with locally convex topologies consistent with the duality (Mackey's Theorem).

PROPOSITION (1): *Let E be a separated locally convex space. The topological dual E', endowed with its equicontinuous bornology, is a complete convex bornological space.*

Proof: If we give E' the weak topology $\sigma(E',E)$, then an equicontinuous set $H \subset E'$ is bounded for such a topology and it suffices to show that H is sequentially complete for $\sigma(E',E)$ (*cf.* Proposition (1) of Section 3:1). Now if (x_n') is a Cauchy sequence in H for $\sigma(E',E)$, then for each $x \in E$, $\langle x,x_n' \rangle$ is a Cauchy sequence in $\mathbb{K}$ and, therefore, converges to an element $u(x)$ of $\mathbb{K}$. The map $u:x \to u(x)$ is clearly a linear map of E into $\mathbb{K}$. By Theorem (2) of Section 5:1, we may assume H to be of the form $H = V^\circ$, with V a neighbourhood of 0 in E. Then $|\langle x,x_n' \rangle| \leqslant 1$ for all $x \in V$ and for all n and hence, passing to the limit, $|u(x)| \leqslant 1$ for all $x \in V$, which implies that u is continuous (and that it belongs to $V^\circ = H$). It is now clear that the sequence (x_n') converges to u for $\sigma(E',E)$.

The completeness of the equicontinuous bornology yields, in a natural way, the existence of a completely bornological topology associated with such a bornology, according to the general scheme set out in Section 4:3. Hence the following Definition:

DEFINITION (1): *Let E be a separated locally convex space and let E' be its dual equipped with the equicontinuous bornology. The space ${}^tE'$ is called the* ULTRA-STRONG DUAL *of E and its topology is called the* ULTRA-STRONG TOPOLOGY *of E'.*

Thus the disks of E' which absorb the equicontinuous sets form a base of neighbourhoods of 0 for the ultra-strong topology. Since equicontinuous sets are strongly bounded, the ultra-strong topology is always finer than the strong topology. E' being a complete convex bornological space by Proposition (1), we can now state the following Theorem, also by virtue of the definition of a completely bornological topology.

THEOREM (1): *The ultra-strong dual of a separated locally convex space is a completely bornological space.*

This Theorem is very important, providing practically the only tool for proving that one of the usual topologies on the dual of a locally convex space is bornological; in fact, one shows that this topology coincides with the topology of ${}^tE'$. In this way one establishes, for example, that L. Schwartz's spaces of distributions are completely bornological.

A consequence of Theorem (1) is the following equally important Theorem.

COROLLARY (1): (Mackey's Theorem): *Let E be a separated locally convex space. A subset of E is bounded for the topology $\sigma(E,E')$ if and only if it is bounded for the given topology on E.*

Proof: Since the given topology on E is always finer than the weak topology $\sigma(E,E')$ every set bounded for the former is obviously bounded for the latter. To see the converse, consider a subset A of E bounded for $\sigma(E,E')$; then A° is an absorbent disk in E' which is closed for $\sigma(E',E)$, hence also for the topology of ${}^tE'$, the latter being always finer than the former. Now ${}^tE'$ is completely bornological (Theorem (1)), whence barrelled (Proposition (5) of Section 5:2), hence A° is bornivorous in ${}^tE'$, *a fortiori*, A° absorbs each equicontinuous subset of E'. It follows that $A^{\circ\circ}$ is absorbed by every neighbourhood of 0 in E, therefore $A^{\circ\circ}$ is bounded in E and, *a fortiori*, so is A.

The above Corollary may be stated in the following more general form.

COROLLARY (2): *Let (F,G) be a separated duality. All separated locally convex topologies on F, consistent with this duality, have the same von Neumann bornology, which is the von Neumann bornology of $\sigma(F,G)$.*

5:4 COMPLETENESS OF THE NATURAL TOPOLOGY ON A BORNOLOGICAL DUAL

PROPOSITION (1): *Let E be a regular convex bornological space. The bornological dual $E^\times$, endowed with its natural topology, is a complete locally convex space.*

Proof: Let (u_j) be a Cauchy net in $E^\times$ (Subsection 0·C.4·5); for every neighbourhood V of 0 in $E^\times$ there exists j_0 such that $(u_j - u_{j'}) \in V$ whenever $j,j' \geqslant j_0$. For each $x \in E$, the net $(u_j(x))$ is a Cauchy net in $\mathbb{K}$ and hence converges to an element $u(x) \in \mathbb{K}$. This defines a linear functional $u:x \to u(x)$ on E and it suffices to show that u is bounded. Now if A is bounded in E, A° is a neighbourhood of 0 in $E^\times$ and hence $(u_j - u_{j'}) \in A^\circ$ for j,j' 'large enough', or, equivalently, $\sup_{x\in A}|u_j(x) - u_{j'}(x)| \leqslant 1$. Passing to the limit on j' we obtain $\sup_{x\in A}|u_j(x) - u(x)| \leqslant 1$ and, since $u_j(A)$ is bounded, we deduce that $u(A)$ is also bounded.

COROLLARY: *The strong dual of a separated bornological locally convex space is complete.*

Proof: If E is a bornological locally convex space, then $E'_\beta = ({}^bE)^\times$ algebraically and topologically.

REMARK: In practice, one appeals to the above Corollary in order to prove the completeness of the strong duals most frequently encountered in Analysis.

5:5 EXTERNAL DUALITY BETWEEN BOUNDED AND CONTINUOUS LINEAR MAPS: DUAL MAPS

5:5·1 Definition of a Dual Map

Let (F,G) and (F_1,G_1) be dualities and let $u:F \to F_1$ be a linear map. For every $y^* \in F_1^*$ the map $y^* \circ u : x \to \langle u(x), y^*\rangle$ is a linear functional on F, denoted by $u^*(y^*)$. Thus by definition we have:

$$\langle u(x), y^*\rangle = \langle x, u^*(y^*)\rangle,$$

for all $x \in F$ and $y^* \in F_1^*$. The map $u^*: y^* \to u^*(y^*)$ is a linear map of the algebraic dual F_1^* of F_1 into the algebraic dual F^* of F and is called the *ALGEBRAIC DUAL (MAP)* of u.

Suppose that the above dualities are separated. Since G_1 (resp. G) may be regarded as a subspace of F_1^* (resp. F^*), the restriction of u^* to G_1 is a linear map of G_1 *into* F^*. The following Proposition tells us when this restriction takes its values in G.

PROPOSITION (1): *$u^*(G_1) \subset G$ if and only if u is continuous for the weak topologies $\sigma(F,G)$ and $\sigma(F_1,G_1)$.*

Proof: If $u^*(G_1) \subset G$, then $u^*(y^*) \in G$ for all $y^* \in G_1$ and the map $x \to \langle u(x),y^*\rangle = \langle x,u^*(y^*)\rangle$ is continuous for $\sigma(F,G)$. Since this holds for all $y^* \in G_1$, the map $x \to u(x)$ is continuous for $\sigma(F,G)$ and $\sigma(F_1,G_1)$. Conversely, if u is continuous for these topologies, then the map $x \to \langle u(x),y^*\rangle = \langle x,u^*(y^*)\rangle$ is continuous for $\sigma(F,G)$ and hence $u^*(y^*) \in G$ (Proposition (2) of Section 5:0).

If $u^*(G_1) \subset G$, we denote by u' the restriction of u^* to G_1. u' is a linear map of G_1 into G called the *DUAL MAP of u (with respect to the given dualities).*

PROPOSITION (2): *Let the linear map $u:F \to F_1$ be continuous for $\sigma(F,G)$ and $\sigma(F_1,G_1)$. Then the dual $u':G_1 \to G$ of u is continuous for $\sigma(G_1,F_1)$ and $\sigma(G,F)$, and $u'' = u$.*

Proof: Since:

$$\langle u(x),y^*\rangle = \langle x,u^*(y^*)\rangle = \langle x,u'(y^*)\rangle,$$

for all $x \in F$ and all $y^* \in G_1$, we see, as in the proof of Proposition (1), that u' is continuous for the appropriate weak topologies. Moreoever, interchanging the rôles of F,F_1 and G_1,G in Proposition (1), we obtain $u'' = u$.

5:5·2 Elementary Properties of Dual Maps

PROPOSITION (3): *Let (F,G) and (F_1,G_1) be separated dualities, let $u:F \to F_1$ be a weakly continuous linear map with dual u' and let A,B be subsets of F,F_1 respectively. Then:*

(a): $(u(A))^\circ = (u')^{-1}(A^\circ)$;

(b): *If $u(A) \subset B$, then $u'(B^\circ) \subset A^\circ$.*

Proof: (a): $(u(A))^\circ = \{y^* \in G_1; |\langle u(x),y^*\rangle| \leqslant 1 \text{ for all } x \in A\} = \{y^* \in G_1; |\langle x,u'(y^*)\rangle| \leqslant 1 \text{ for all } x \in A\} = (u')^{-1}(A^\circ)$.

(b): $u(A) \subset B$ implies $B^\circ \subset (u(A))^\circ = (u')^{-1}(A^\circ)$, which implies $u'(B^\circ) \subset A^\circ$.

COROLLARY (1): $\ker u' = (u(F))^\circ$.

Proof: By virtue of Proposition (3)(a) we have:

$$\ker u' = (u')^{-1}(0) = (u')^{-1}(F^\circ) = u(F)^\circ.$$

COROLLARY (2): *u' is injective if and only if $u(F)$ is dense in F_1 for $\sigma(F_1,G_1)$.*

Proof: If u' is injective, then $(u(F))^\circ = \{0\}$ by Corollary (1), hence $(u(F))^{\circ\circ} = \{0\}^\circ = F_1$. But, by the Bipolar Theorem, $(u(F))^{\circ\circ} = \overline{u(F)}$, the closure of $u(F)$ for $\sigma(F_1,G_1)$, and hence $u(F)$ is dense in F_1 for $\sigma(F_1,G_1)$. Conversely, if this is true, then $\ker u' = (u(F))^\circ = (\overline{u(F)})^\circ = F_1^\circ = \{0\}$ and u' is injective.

5:5·3 External Duality between Bounded and Continuous Linear Maps

THEOREM (1): *Let (F,G) and (F_1,G_1) be separated dualities and let u be a weakly continuous linear map of F into F_1 with dual map u'. Let $\mathcal{B}$ (resp. $\mathcal{B}_1$) be a convex bornology on F (resp. F_1) compatible with the topology $\sigma(F,G)$ (resp. $\sigma(F_1,G_1)$) and let $\mathcal{B}^\circ$ (resp. $\mathcal{B}_1^\circ$) be the $\mathcal{B}$-topology on G (resp. the $\mathcal{B}_1$-topology on G_1). Then:*

(a): *If u is bounded from $(F,\mathcal{B})$ into $(F,\mathcal{B}_1)$, u' is continuous from $(G_1,\mathcal{B}_1^\circ)$ into $(G,\mathcal{B}^\circ)$;*

(b): *Suppose that the members of $\mathcal{B}$ (resp. $\mathcal{B}_1$) are closed for $\sigma(F,G)$ (resp. $\sigma(F_1,G_1)$). If u' is continuous from $(G_1,\mathcal{B}_1^\circ)$ into $(G,\mathcal{B}^\circ)$, u is bounded from $(F,\mathcal{B})$ into $(F_1,\mathcal{B}_1)$.*

Proof: The relation $u(A) \subset B$ implies $u'(B^\circ) \subset A^\circ$ and hence (a). For (b), consider $A \in \mathcal{B}$; by virtue of the continuity of u' there exists $B \in \mathcal{B}_1$ such that $u'(B^\circ) \subset A^\circ$. Now the Bipolar Theorem and Proposition (3) imply:

$$A = A^{\circ\circ} \subset (u'(B^\circ))^\circ = (u'')^{-1}(B^{\circ\circ}) = u^{-1}(B^{\circ\circ}) = u^{-1}(B),$$

hence $u(A) \subset B$ and, consequently, u is bounded.

5:5·4 Particular Cases

We consider the two most important particular cases, which occur when F and F_1 are separated locally convex spaces (resp. regular convex bornological spaces), G and G_1 are their topological (resp. bornological) duals and $u:F \to F_1$ is a continuous (resp. bounded) linear map.

If $u:F \to F_1$ is a continuous linear map between two locally convex spaces, then u is continuous from $\sigma(F,F')$ to $\sigma(F_1,F_1')$. In fact, for every $y_1' \in F_1'$ the linear map $x \to \langle u(x),y_1'\rangle$ is continuous on F, hence continuous for $\sigma(F,F')$, since the topology $\sigma(F,F')$ is consistent with the duality between F and F'; thus u is continuous for the weak topologies. We can then form the dual map u': $F_1' \to F'$, which is both weakly and strongly continuous (Theorem (1)), i.e. continuous when F_1' and F' are given either the topologies $\sigma(F_1',F_1)$ and $\sigma(F',F)$ or the topologies $\beta(F_1',F_1)$ and $\beta(F',F)$.

Suppose now that F and F_1 are regular convex bornological spaces and that $u:F \to F_1$ is a bounded linear map; u is continuous from ${}^tF$ to ${}^tF_1$, hence, by the above, continuous from $\sigma(F,F^\times)$ to $\sigma(F_1,F_1^\times)$, since $F^\times = ({}^tF)'$ and $F_1^\times = ({}^tF_1)$. Now every bounded subset of F (resp. F_1) is bounded for $\sigma(F,F^\times)$ (resp. $\sigma(F_1,F_1^\times)$)

and Theorem (1) implies that the dual $u':F_1^\times \to F^\times$ is continuous for the natural topologies on $F_1^\times$ and $F^\times$.

5:5·5 Boundedness of the Dual Map

PROPOSITION (4): *Let F and F_1 be separated locally convex spaces with duals F' and F_1' respectively, and let $u:F \to F_1$ be a weakly continuous linear map (i.e. continuous for $\sigma(F,F')$ and $\sigma(F_1,F_1')$). Then u is continuous for the given topologies on F and F_1 if and only if its dual $u':F_1' \to F'$ is bounded for the equicontinuous bornologies of F_1' and F'.*

Proof: Denote by $\mathfrak{K}$ (resp. $\mathfrak{K}_1$) the equicontinuous bornology of F' (resp. F_1'); this bornology has a base consisting of polars of neighbourhoods of 0 in F (resp. F_1) (Theorem (2) of Section 5:1). Since the given topology on F (resp. F_1) is the $\mathfrak{K}$-topology (resp. the $\mathfrak{K}_1$-topology) (Theorem (3) of Section 5:1), we may apply Theorem (1) to conclude that $u = u''$ is continuous from F to F_1 if and only if $u':F_1' \to F'$ is bounded.

PROPOSITION (5): *Let F and F_1 be regular convex bornological spaces with bornological duals $F^\times$ and $F_1^\times$ respectively, and let $u:F \to F_1$ be a bounded linear map. Then the dual $u':F_1^\times \to F^\times$ is bounded when $F_1^\times$ and $F^\times$ are given their natural bornologies.*

Proof: The natural bornology on a bornolgical dual consists of all subsets that are equibounded on each bounded set. Let H_1 be a subset of $F_1^\times$, bounded for the natural bornology of $F_1^\times$; we show that $u'(H_1)$ is bounded for the natural bornology of $F^\times$. If A is a bounded set in F, then $u(A)$ is bounded in F_1 and hence $H_1(u(A))$ is bounded in $\mathbb{K}$. This concludes the proof, since $u'(H_1)(A) = H_1(u(A))$.

'TOPOLOGY—BORNOLOGY': EXTERNAL DUALITY

II — WEAKLY COMPACT BORNOLOGIES AND REFLEXIVITY

From the point of view of the applications, an important class of spaces is the class of (locally convex or convex bornological) spaces whose bounded sets are weakly relatively compact. In such spaces one can extract, under suitable conditions, a weakly convergent subsequence from every bounded sequence, and even a 'strongly convergent' one if certain compactness hypotheses are satisfied (in a sense to be made precise in the next Chapter).

The object of the present Chapter is to characterise those spaces whose bornologies are weakly compact. This is found to be equivalent to *the problem of the representation of a given space as the space of 'continuous or bounded' linear functionals on its dual*, which is what is meant by 'reflexivity'. Our approach to the reflexivity theory differs from the classical one in several respects.

Starting with a separated locally convex space E with dual E', there are two natural ways of relating E to a space of linear functionals on E'. The first, which is the only one treated in the classical literature, consists in giving E' the strong topology and in considering the space E'' of continuous linear functionals on E'_β, whilst in the second we give E' its equicontinuous bornology and study the space $(E')^\times$ of bounded linear functionals on E'. In the first case E'', being a topological dual, is naturally endowed with its equicontinuous bornology and we say that E is *reflexive* if $E = E''$ algebraically, hence bornologically. In the second case $(E')^\times$, being a bornological dual, is canonically equipped with a topology (its natural topology) and we say that E is *completely reflexive* if $E = (E')^\times$ algebraically, hence topologically; such spaces are studied in Section 6:4. In the classical theory only the first case is considered and E is called *semi-reflexive* if $E = E''$ algebraically. But this presentation hides the fact that if the above algebraic identity if of interest, it is so precisely because of the underlying bornological identity, from which the weak compactness of bounded subsets of E originates.

Since, *a priori*, there is no reason to consider the strong topology on E'' (unless we wished to study the reflexivity of E' in our sense), the classical notion of 'reflexivity' will not appear here. For the applications, only complete reflexivity, more powerful than classical reflexivity, will be considered.

Naturally, a scheme dual to the one presented above for locally convex spaces is established, in Section 6:3, to characterise convex bornological spaces with weakly compact bornologies.

Section 6:2 gives the Mackey-Arens Theorem in its true form, i.e. as a characterisation of these bornologies that are compatible with a topological duality. Obviously, our statement of this Theorem will be different from the classical ones.

The basic result for all questions relating to weak compactness is the weak compactness of equicontinuous sets, which is established in Section 6:1.

6:1 WEAK COMPACTNESS OF EQUICONTINUOUS SETS

THEOREM (1): *Let E be a separated locally convex space. Every equicontinuous subset of E' is relatively compact for the topology $\sigma(E',E)$.*

Proof: Since every equicontinuous subset of E is contained in the polar V° of a neighbourhood V of 0 in E (Theorem (2) of Section 5:1), it suffices to show that the set $H = V^\circ$ is compact for $\sigma(E',E)$. The dual E', endowed with $\sigma(E',E)$, is a topological sub space of the product space $\mathbb{K}^E$, the canonical injection being:

$$x' \to (\langle x,x'\rangle)_{x\in E}.$$

For every $x \in E$ the set $H(x)$ is bounded in $\mathbb{K}$ (Proposition (3) of Section 5:2) and hence its closure $B_x = \overline{H(x)}$ is compact in $\mathbb{K}$; by Tychonov's Theorem (Section 0·B; see L. Schwartz [*1*]) the set $\prod_{x\in E} B_x$ is compact in $\mathbb{K}^E$ and, since it contains H, H is relatively compact in $\mathbb{K}^E$. It is enough to show that H is closed in $\mathbb{K}^E$, for then H is compact in $\mathbb{K}^E$ and, being contained in E', is compact in E' for $\sigma(E',E)$. Let (u_j) be a net of elements of H such that for every $x \in E$, $(u_j(x))$ converges to an element $\ell(x)$ in $\mathbb{K}$. The map u defined by $u(x) = \ell(x)$ is clearly linear from E into $\mathbb{K}$. Since $|u_j(x)| \leqslant 1$ for all $x \in V$ and all j, we must have $|u(x)| \leqslant 1$ for all $x \in V$ and hence u is continuous and belongs to $V^\circ = H$. Thus H is closed in $\mathbb{K}^E$ and the Theorem follows.

From the weak compactness of equicontinuous sets we can deduce compactness for finer topologies via the following Lemma.

LEMMA (1): *Let E be a separated locally convex space with dual E'. On each equicontinuous set $H \subset E'$ the weak topology $\sigma(E',E)$ and the topology of precompact convergence coincide.*

Proof: Since a finite subset of E is precompact, $\sigma(E',E)$ is coarser than the topology of precompact convergence and hence it suffices to prove that it is finer on each equicontinuous subset H of E'. We have to show that for every $x_0' \in H$ and precompact set $A \subset E$, there exists a finite set $B \subset E$ such that:

$$(x_0' + B^\circ) \cap H \subset (x_0' + A^\circ) \cap H. \tag{1}$$

The set $H - x_0'$ being equicontinuous, its polar U is a neighbourhood of 0 in E, and if we put $V = \frac{1}{2}U$, we have:

$$\sup_{\substack{x' \in H \\ x \in V}} |\langle x' - x_0', x\rangle| \leqslant \tfrac{1}{2}.$$

In view of the precompactness of A there are finitely many points $a_1, \ldots, a_n$ in E such that $A \subset \bigcup_{i=1}^{n} (a_i + V)$ and hence each $x \in A$ is of the form $x = a_i + y$, $y \in V$. Putting $x_i = 2a_i$ and $B = \{x_1, \ldots, x_n\}$ we show that (1) holds. Let $x' \in H$ be such that $(x' - x_0') \in B^\circ$, i.e. $\sup_{1 \leqslant i \leqslant n} |\langle x' - x_0', x_i\rangle| \leqslant 1$; we have:

$$\begin{aligned}\sup_{x \in A} |\langle x' - x_0', x\rangle| &\leqslant \sup_{1\leqslant i\leqslant n} |\langle x' - x_0', a_i\rangle| + \sup_{y \in V} |\langle x' - x_0', y\rangle| \\ &\leqslant \tfrac{1}{2} \sup_{1\leqslant i\leqslant n} |\langle x' - x_0', x_i\rangle| + \sup_{y \in V} |\langle x' - x_0', y\rangle| \leqslant 1.\end{aligned}$$

and therefore $x' \in x_0' + A^\circ$.

Theorem (1) and Lemma (1) imply the following result, which will be used in Chapter VII.

PROPOSITION (1): *Let E be a separated locally convex space. Every weakly closed equicontinuous subset of E' is compact for the topology of precompact convergence.*

COROLLARY: *Let E be a separated locally convex space. Every weakly closed equicontinuous subset of E' is compact for the topology of compact convergence.*

Proof: In fact, since a compact subset of E is precompact, the topology of compact convergence is coarser than the topology of precompact convergence.

6:2 THE BORNOLOGY OF WEAKLY COMPACT DISKS AND THE MACKEY-ARENS THEOREM

6:2·1

DEFINITION (1): *Let (F,G) be a separated duality. A convex bornology $\mathcal{B}$ on G is said to be a* CONVEX BORNOLOGY COMPATIBLE WITH THE DUALITY *between F and G if it satisfies the following conditions:*

(i): *$\mathcal{B}$ is compatible with the topology* $\sigma(G,F)$;

(ii): *The $\mathcal{B}$-topology on F is compatible with* (F,G) *(cf. Definition* (2) *of Section* 5:1*)*.

For example, if $F = E$ is a separated locally convex space and $G = E'$ is its topological dual, then the given topology on E, and hence the *equicontinuous bornology* on E', are always compatible with (E,E'). Another example of a bornology which is compatible with the topological duality between two arbitrary spaces F and G is afforded by the *finite-dimensional bornology* of G: in fact, finite-dimensional bounded sets are bounded for every vector bornology on G, in particular, for $\sigma(G,F)$; moreover, the finite-dimensional bornology of G yields, by polarity, the topology $\sigma(F,G)$ on F.

The reader will notice that the above examples of '*compatible bornologies*' have bases consisting of weakly compact disks (see Section 6:1 for the weak compactness of equicontinuous sets). We shall presently show that this is not due to chance; indeed, an important theorem (the Mackey-Arens Theorem) asserts precisely that weak compactness is a necessary and sufficient condition for a bornology to be compatible with a topological duality.

6:2·2 The Mackey-Arens Theorem

THEOREM (1): (Mackey-Arens Theorem): *Let (F,G) be a separated duality. A convex bornology on G is compatible with (F,G) if and only if it has a base consisting of disks that are relatively compact for* $\sigma(G,F)$.

Proof: Necessity: Let $\mathcal{A}$ be a convex bornology on G, compatible with (F,G); we denote by $\mathcal{B}$ the convex bornology on G having as a base $\mathcal{B}_0$ the closures for $\sigma(G,F)$ of disked members of $\mathcal{A}$. On F the $\mathcal{A}$-topology and the $\mathcal{B}$-topology are the same (Remark (1) of Section 5:1), hence it suffices to show that the members of $\mathcal{B}_0$ are compact for $\sigma(G,F)$. Let $\mathcal{T}$ be the $\mathcal{B}$-topology on F; we put $E = (F,\mathcal{T})$ so that $E' = G$. Since $\{B^\circ; B \in \mathcal{B}_0\}$ is a base of neighbourhoods of 0 in E, the family $\{B^{\circ\circ}; B \in \mathcal{B}_0\}$ is a base for the equicontinuous bornology on E'. But $B^{\circ\circ} = B$, since B is closed for $\sigma(G,F)$ (Bipolar Theorem), hence $\mathcal{B}$ is the equicontinuous bornology of E'. Now weakly closed equicontinuous sets are weakly compact (Theorem (1) of Section 6:1) and the necessity follows.

Sufficiency: Let $\mathcal{A}$ be a convex bornology on G with a base of relatively compact disks for $\sigma(G,F)$ and let $\mathcal{B}$ be the convex bornology on G having as a base $\mathcal{B}_0$ the closures for $\sigma(G,F)$ of members of $\mathcal{A}$. Clearly the elements of $\mathcal{B}$ remain bounded for $\sigma(G,F)$. Denote by $\mathcal{T}$ the $\mathcal{B}$-topology on F, which is also the $\mathcal{A}$-topology; we have to show that the dual of $(F,\mathcal{T})$ is G. Put $E = (F,\mathcal{T})$.

(a): First of all, $G \subset E'$; in fact, a finite-dimensional bound[ed] subset of G which is closed for $\sigma(G,F)$ is compact for $\sigma(G,F)$, bei[ng] closed and bounded in a finite-dimensional subspace of G. Hence, by polarity, $\mathcal{T}$ if finer than $\sigma(F,G)$ and every linear functional o[n] F, continuous for $\sigma(F,G)$, is also continuous for $\mathcal{T}$. But every linear functional on F which is continuous for $\sigma(F,G)$ is uniquely given by an element of G and conversely; therefore, G is contained in E'.

(b): Next, let F^* be the algebraic dual of F. If we identify G with a subspace of F^*, the topology $\sigma(G,F)$ coincides with the topology induced on G by $\sigma(F^*,F)$. Since every $B \in \mathcal{B}_0$ is compact for $\sigma(G,F)$, its image in F^* under the canonical embedding $G \to F^*$ is compact for $\sigma(F^*,F)$. But B is a disk, hence $(B^\circ)^\circ_{F^*} = B$, where $(\cdot)^\circ_{F^*}$ denotes the polar in F^* with respect to the duality (F,F^*).

(c): The family $\{B^\circ; B \in \mathcal{B}_0\}$ is a base of neighbourhoods of 0 in E, whence $\{(B^\circ)^\circ_{E'}; B \in \mathcal{B}_0\}$ (polars in E') is a base for the equicontinuous bornology of E' and $E' = \bigcup_{B \in \mathcal{B}_0} (B^\circ)^\circ_{E'}$. However, $(B^\circ)^\circ_{E'} = (B^\circ)^\circ_{F^*} \cap E'$ and, by (b), $(B^\circ)^\circ_{F^*} = B \subset G \subset E'$, so that $(B^\circ)^\circ_{E'} = B$ and, consequently, $E' = \bigcup_{B \in \mathcal{B}_0} B = G$.

COROLLARY (1): *Let (F,G) be a separated duality. The bornology of weakly compact disks of G is the finest convex bornology on G compatible with (F,G).*

COROLLARY (2): *Let (F,G) be a separated duality. A separated locally convex topology $\mathcal{T}$ on F is compatible with (F,G) if and only if $\mathcal{T}$ is a $\mathcal{B}$-topology for a convex bornology $\mathcal{B}$ on G with a base of weakly compact disks.*

Proof: The necessity has already been seen (weak compactness of equicontinuous sets), whilst the sufficiency follows from Theorem (1).

6:2·3 The Mackey Topology

DEFINITION (2): *The MACKEY TOPOLOGY on F RELATIVE TO THE DUALITY (F,G), denoted by $\tau(F,G)$, is the $\mathcal{B}$-topology on F when $\mathcal{B}$ is the bornology on G with a base consisting of all disks that are compact for $\sigma(G,F)$*

In view of the Mackey-Arens Theorem, such a topology is compatible with the duality between F and G and is the finest separated locally convex topology on F with this property. Since every equicontinuous subset of the dual E' of a separated locally convex space is compact for $\sigma(E',E)$, we see, by polarity, that the Mackey topology $\tau(E,E')$ on E is finer than the given topology of E. A separated locally convex space whose topology coincides with the Mackey topology is called a *MACKEY SPACE*, and can be characterised as follows:

PROPOSITION (1): *A separated locally convex space E is a Mackey space if and only if every weakly compact disk of E' is equicontinuous.*

Proof: Let $\mathcal{K}$ be the equicontinuous bornology of E' and let $\mathcal{C}$ be the bornology generated by the disks of E' that are compact for $\sigma(E',E)$. The given topology on E is the $\mathcal{K}$-topology whilst the Mackey topology is the $\mathcal{C}$-topology and the two topologies agree if $\mathcal{K} = \mathcal{C}$, hence the necessity. Conversely, if the two topologies coincide, then $\mathcal{K}^\circ = \mathcal{C}^\circ$ and hence $\mathcal{K}^{\circ\circ} = \mathcal{C}^{\circ\circ}$. But $\mathcal{K}^{\circ\circ} = \mathcal{K}$ and $\mathcal{C}^{\circ\circ} = \mathcal{C}$, since $\mathcal{C}$ has a base of disks, and, therefore, $\mathcal{K} = \mathcal{C}$.

The most important example of a Mackey space is given by the following Proposition.

PROPOSITION (2): *Every separated bornological locally convex space E is a Mackey space. In particular, every metrizable locally convex space is a Mackey space.*

The second assertion follows from the first (Proposition (3) of Section 4:1). To establish the latter, we shall give two proofs.

First Proof: The given topology $\mathcal{T}$ on E is coarser than the Mackey topology $\tau(E,E')$. Both topologies are compatible with the duality (E,E') and hence have the same von Neumann bornology (Corollary (2) to Theorem (1) of Section 5:3). It follows that the identity $(E,\mathcal{T}) \to (E,\tau(E,E'))$ is bounded, hence continuous, since E is bornological.

Second Proof: We shall show that a separated locally convex space E is a Mackey space if every strongly bounded subset of E' is equicontinuous; this will imply Proposition (2) by the Corollary to Proposition (6) of Section 5:2. Now the assertion to be proved follows from:

LEMMA (1): *Let E be a separated locally convex space. Every weakly compact disk of E' is strongly bounded.*

Proof: Let $\mathcal{C}$ be the convex bornology on E' generated by the family of disks that are compact for $\sigma(E',E)$; $\mathcal{C}$ is a complete bornology (Corollary to Proposition (1) of Section 3:1). Denote by F the complete convex bornological space $(E',\mathcal{C})$. The locally convex space ${}^tF$ is completely bornological by definition, hence barrelled (Proposition (5) of Section 5:2). We have to show that for every bounded set $A \subset E$, A° is bornivorous in F, i.e. a neighbourhood of 0 in ${}^tF$ (since A° is a disk). Since ${}^tF$ is barrelled, it suffices to prove that the disk A° is absorbent in $F = E'$ and closed in ${}^tF$. Now A° is absorbent, because A is bounded in E, hence bounded for $\sigma(E,E')$; moreover, A° is closed for $\sigma(E',E)$, hence closed in the finer topology of ${}^tF$.

6:3 WEAKLY COMPACT BORNOLOGIES: REFLEXIVITY

In this Section we characterise those (locally convex or convex bornological) spaces whose bornologies have bases consisting of weakly compact disks. We shall have to distinguish between locally convex spaces and convex bornological spaces since, if E belongs to the latter class, the natural weak topology on E is the topology $\sigma(E,E^\times)$, with $E^\times$ the bornological dual of E, whilst if E belongs to the former class, then the natural weak topology on E is the topology $\sigma(E,E')$, with E' the topological dual of E.

6:3˙1 Convex Bornological Spaces with a Weakly Compact Bornology

6:3˙1(a) *Bidual Space and Reflexivity*

Let E be a regular convex bornological space, let $E^\times$ be its bornological dual with the natural topology and let $(E^\times)'$ be the topological dual of $E^\times$ with its equicontinuous bornology. The space $(E^\times)'$ is called the *BIDUAL of E*.

With every $x \in E$ we can associate the linear functional:

$$u_x : y \to \langle x,y \rangle \qquad \text{for all } y \in E^\times.$$

u_x is obviously continuous on $E^\times$ (even continuous for $\sigma(E^\times,E)$) and hence belongs to $(E^\times)'$. Since $E^\times$ separates E, the map $x \to u_x$ of E into $(E^\times)'$ is a linear injection called the *CANONCIAL EMBEDDING of E into* $(E^\times)'$. Under this map, E can be identified with a vector subspace of $(E^\times)'$ but, in general, E cannot be regarded as a bornological subspace of $(E^\times)'$. Those convex bornological spaces for which this is true will be characterised later.

A regular convex bornological space E is said to be REFLEXIVE if $E = (E^\times)'$ *algebraically and bornologically.*

We have:

6:3˙1(b)

PROPOSITION (1): *Let E be a regular convex bornological space. Then E has a base of disks compact for* $\sigma(E,E^\times)$ *if and only if E is reflexive.*

Proof: Necessity: Let $\mathcal{B}$ be a base of the bornology of E consisting of disks compact for $\sigma(E,E^\times)$ and let $B \in \mathcal{B}$. Since $\sigma(E,E^\times)$ is the topology induced by $\sigma((E^\times)',E^\times)$ on E, B is compact in $(E^\times)'$ for $\sigma((E^\times)',E^\times)$, hence closed for this topology and, by the Bipolar Theorem, B coincides with its bipolar $B^{\circ\circ}$ in $(E^\times)'$. It follows that:

$$(E^\times)' = \bigcup_{B \in \mathcal{B}} B^{\circ\circ} = \bigcup_{B \in \mathcal{B}} B = E,$$

and, therefore, E is reflexive.

Sufficiency: If $E = (E^\times)'$ algebraically and bornologically, then the bounded subsets of E are the same as the equicontinuous sets in $(E^\times)'$ and the latter are relatively compact for $\sigma((E^\times)',E^\times) = \sigma(E,E^\times)$.

6:3˙1(c) *Polar Convex Bornological Spaces*

A regular convex bornological space which is a bornological subspace of its bidual is called POLAR. Such spaces can be characterised as follows:

PROPOSITION (2): *Let E be a regular convex bornological space. The following assertions are equivalent:*

(i): *E is a bornological subspace of $(E^{\times})'$;*

(ii): *For every bounded subset A of E, the bipolar of A in E with respect to the duality $(E,E^{\times})$ is again bounded in E;*

(iii): *E has a base of bounded sets which are closed in ${}^{t}E$.*

Proof: The equivalence of (i) and (ii) follows from the fact that the bornology of $(E^{\times})'$ has as a base the bipolars in $(E^{\times})'$ of all bounded subsets of E and hence it induces a bornology on E having as a base the bipolars in E of the bounded subsets of E.

(ii) => (iii): In fact, for every bounded subset A of E, the bipolar of A in E with respect to $(E,E^{\times})$ is closed for $\sigma(E,E^{\times}) = \sigma({}^{t}E,({}^{t}E)')$, hence closed in ${}^{t}E$ (Corollary (2) to Theorem (2) and Remark (1) of Section 5:0). Since such bipolars are bounded, the form a base for the bornology of E and (iii) follows.

(iii) => (i): Similarly, because every bounded disk $A \subset E$, whic is closed in ${}^{t}E$, is closed for $\sigma(E,E^{\times})$ and hence coincides with it bipolar, the implication follows.

EXAMPLE (1): *If E is a separated locally convex space, then ${}^{b}E$ is polar:* In fact, since the topology of ${}^{tb}E$ is finer than that of E, ${}^{b}E$ is regular. Moreover, ${}^{b}E$ has a base of bounde disks which are closed in E, hence closed for $\sigma(E,E')$ and, *a fort iori*, closed for the finer topology $\sigma({}^{b}E,({}^{b}E)^{\times})$.

EXAMPLE (2): *If E is a separated locally convex space, then the space E', endowed with its equicontinuous bornology, is polar:* Since E separates E', $(E')^{\times}$ separates E' and E' is regular. Also, E' has a base of disks which are equicontinuous, hence compact for $\sigma(E',E)$, hence closed for $\sigma(E',E)$ and, therefor closed for the finer topology $\sigma(E',(E')^{\times})$.

6:3·2 Locally Convex Spaces with a Weakly Compact von Neumann Bornology

6:3·2(a) *Bidual Space and Reflexivity*

Let E be a separated locally convex space, let E' be its topo logical dual with the strong topology and let E'' be the topologic dual of E', endowed with the equicontinuous bornology. The space E'' is called the *BIDUAL* of E and is, by definition, a convex born ological space.

As in Subsection 6:3·1, the map which associates with every $x \in E$ the linear functional $u_x : x' \to \langle x,x' \rangle$ on E', is a linear injection of E into E'' called the *CANONCIAL EMBEDDING of E into E''* Via this map we can identify E with a subspace of E''. Since the bornology induced by E'' on E has as a base precisely the intersections with E of the bipolars in E'' of the closed bounded disk of E, such a bornology coincides with the von Neumann bornology of E, and so we see that *the space ${}^{b}E$ is automatically a bornolo ical subspace of E''.*

A separated locally convex space E is said to be REFLEXIVE if

$E = E''$ *(algebraic, hence bornological, duality).*

It should be noted that the term 'reflexive' is not used here in the sense of Bourbaki [3] (see also the introduction of this Chapter).

The above definition shows immediately that a separated reflexive locally convex space has a von Neumann bornology with a base of disks compact for $\sigma(E,E') = \sigma(E'',E')$. The converse is also true.

6:3·2(b)

PROPOSITION (3): *Let E be a separated locally convex space. The von Neumann bornology of E has a base of weakly compact disks if and only if E is reflexive.*

Proof: The sufficiency having already been seen, we prove the necessity. The argument is similar to the one used in the proof of Proposition (1) (Necessity). Let $\mathcal{B}$ be a base of the bornology of E consisting of disks compact for $\sigma(E,E')$ and let $B \in \mathcal{B}$. Since $\sigma(E,E')$ is the topology induced by $\sigma(E'',E')$ on E, B is compact in E'' for $\sigma(E'',E')$, hence closed for this topology, hence $B = B^{\circ\circ}$ (the bipolar being taken in E''). Thus $E'' = \bigcup_{B\in\mathcal{B}} B^{\circ\circ} = \bigcup_{B\in\mathcal{B}} B = E$.

6:4 COMPLETELY REFLEXIVE LOCALLY CONVEX SPACES

DEFINITIONS: *Let E be a separated locally convex space and let E' be its dual with the equicontinuous bornology.*

(a): *The space $(E')^{\times}$, endowed with its natural topology, is called the* BORNOLOGICAL BIDUAL *of E;*

(b): *E is said to be* COMPLETELY REFLEXIVE *if $E = (E')^{\times}$ algebraically and topologically.*

Note that $E'' \subset (E')^{\times}$, since a continuous linear functional on E'_{β} is bounded on strongly bounded sets, hence on equicontinuous sets (Proposition (2) of Section 5:2). Consequently, *every completely reflexive locally convex space is reflexive*, but the converse if false in general (see Exercise 6·E.3).

The interest of completely reflexive spaces rests essentially on the following Theorem.

THEOREM (1): *Let E be a completely reflexive locally convex space. Then:*

(i): *E is complete;*

(ii): *Every closed bounded subset of E is weakly compact;*

(iii): *The strong dual of E is completely bornological.*

For the proof of this Theorem we need the following two Lemmas.

LEMMA (1): *Let E be a separated locally convex space. Then:*

$$^{b}((E')^{\times}) = (^{t}E')',$$

algebraically and bornologically, where E' and $({}^tE')'$ carry their equicontinuous bornologies and $(E')^\times$ its natural topology.

Proof: It is clear that $(E')^\times = ({}^tE')'$ algebraically. Let H be an equicontinuous subset of $({}^tE')'$; we have to show that H is bounded in ${}^b((E')^\times)$, i.e. equibounded on each equicontinuous subset of E' (Proposition (1) of Section 5:2). Now if A is an equicontinuous set in E', then A is bounded in ${}^{bt}E'$ and hence $H(A)$ is bounded (Proposition (2) of Section 5:2). Conversely, if M is a bounded subset of ${}^b((E')^\times)$, i.e. a subset of $({}^tE')'$ equibounded on each equicontinuous subset of E', then M is equicontinuous in $({}^tE')'$ (Proposition (6) of Section 5:2).

LEMMA (2): *If E is a completely reflexive locally convex space, then the strong and ultra-strong duals of E coincide.*

Proof: Let E' be the topological dual of E; we know that on E' the topology of ${}^tE'$ is always finer than the strong topology. Conversely, let V be a disked neighbourhood of 0 in ${}^tE'$ which is closed for $\sigma({}^tE',({}^tE')') = \sigma(E',(E')^\times) = \sigma(E',E)$. V absorbs the equicontinuous subsets of E', hence its polar V° in E is absorbed by the neighbourhoods of 0 of E and, therefore, V° is bounded in E. Now V° is also the polar of V in $({}^tE')' = E$; moreover, the topology of ${}^tE'$ is the topology of uniform convergence on the sets V° when V runs through a base of neighbourhoods of 0 in ${}^tE'$ (Theorem (3) of Section 5:1). Since such sets V° are bounded in E, the topology of ${}^tE'$ is coarser than the strong topology of E', hence the two topologies must be the same.

Proof of Theorem (1): E being completely reflexive, we have that $E = (E')^\times$ algebraically, hence topologically since $(E')^\times$ is complete (Section 5:4), and (i) follows. Now, since $E = (E')^\times$ topologically, we have that ${}^bE = {}^b((E')^\times)$ bornologically. By Lemma (1) the bornology of ${}^b((E')^\times)$ is the equicontinuous bornology of $({}^tE')'$ and, since every equicontinuous set in $({}^bE')'$ is relatively compact for $\sigma(({}^tE')',{}^tE') = \sigma(E',E^\times) = \sigma(E,E')$, we obtain (ii). Finally, Lemma (2) ensures (iii) by virtue of Theorem (1) of Section 5:3.

CHAPTER VII

COMPACT BORNOLOGIES

The fundamental question in Analysis is the question of convergence. If the bounded subsets of a space are compact for sufficiently fine topologies, then a weakly convergent sequence automatically becomes a 'strongly convergent' one. This is why spaces with compact bounded sets can be regarded as the 'best spaces in Analysis'; they form the object of this Chapter.

The compactness hypotheses that can be imposed on bounded sets are of a diverse nature since, if E is a separated locally convex space, then there are two natural topologies on E: the given topology and the weak topology $\sigma(E,E')$. If we require the bounded subsets of E to be compact for $\sigma(E,E')$, we fall back into the category of reflexive spaces considered in Chapter VI. However, if compactness is assumed in the given topology of E, then we obtain a new class of spaces called *hypo-Montel*, which are studied in Section 7:1. If E is a separated convex bornological space it is natural to consider on E the weak topology $\sigma(E,E^{\times})$ (if E is regular) and the topologies of the spaces E_B spanned by the bounded disks and normed by their gauges. Compactness for $\sigma(E,E^{\times})$ leads to the theory of bornological reflexivity, also treated in Chapter VI, whilst compactness with respect to the spaces E_B yields a new class of spaces and bornologies: the *Schwartz bornologies* and, by duality, the *Schwartz topologies* (Section 7:2) so called after L. Schwartz who was, around 1945, the first to use these important ideas in the particular case of spaces of distributions. Amongst Schwartz bornologies, the *Silva bornologies* (i.e. Schwartz bornologies with a countable base) enjoy very special properties. Their importance was underlined by J.S. Silva, in 1950, in his study of germs of analytic functions, and spaces with Silva bornologies have played an essential rôle in many branches of Functional Analysis ever since. Such spaces are investigated in Section 7:3 and applied in the next Chapter to the solution of partial differential equations.

7:1 HYPO-MONTEL SPACES

7:1·1

DEFINITION: *A separated locally convex space is called* HYPO-MONTEL *if its von Neumann bornology has a base of compact sets. A* MONTEL *space is a locally convex space which is both hypo-Montel and barrelled.*

REMARK (1): This terminology has its origins in properties of bounded subsets of the space $H(\Omega)$ of holomorphic (i.e. differentiable) functions on an open subset Ω of a complex Banach space E, $H(\Omega)$ carrying the topology of uniform convergence on the compact subsets of E. If E has finite dimension, then $H(\Omega)$ is a Montel space (Montel's Theorem), whilst if E has infinite dimension, then $H(\Omega)$ is a hypo-Montel space but not a Montel space. We shall not prove these assertions here, as their proofs appeal to special properties of holomorphic functions.

REMARK (2): *Every hypo-Montel space E is sequentially complete*, since a Cauchy sequence in E is bounded, hence is contained in a compact set and, therefore, converges.

7:1·2 Properties of Hypo-Montel Spaces

THEOREM (1): *Let E be a hypo-Montel space. The given topology on E and the weak topology coincide on each bounded subset of E. Consequently, every weakly convergent sequence in E is also convergent (to the same limit) for the given topology on E.*

Proof: First of all, let us recall an easy result of general topology. Let X be a compact space and let Y be a separated topological space; every continuous bijection $f:X \to Y$ is a homeomorphism. In fact, if A is a closed subset of X, then A is compact in X, hence $f(A)$ is compact in Y and the continuity of f^{-1} is assured. Reverting to the proof of Theorem (1), let B be a bounded subset of E and let $K = \bar{B}$ be the closure of B in E; since E is hypo-Montel, K is compact. Let X be the set K with the topology induced by E and let Y be the set K with the topology induced by $\sigma(E,E')$; the identity $f:X \to Y$ is a continuous bijection, hence a homeomorphism and the first assertion of the Theorem follows. For the second, let (x_n) be a sequence in E which converges to x for $\sigma(E,E')$. The set $A = \{x_n; n \in \mathbb{N}\}$ is bounded for $\sigma(E,E')$ and hence bounded for the topology of E by Mackey's Theorem (Section 5:3). By the first part, the two topologies coincide on A and, therefore, (x_n) converges to x in the topology of E.

7:1·3 A Class of Hypo-Montel Spaces

PROPOSITION (1): *Let F be a barrelled locally convex space, let $\mathcal{B}$ be the bornology of compact disks of F and let $E = F'_c$ be the topological dual of F with the $\mathcal{B}$-topology. Then E is a hypo-Montel space.*

Proof: Let us denote by $\mathcal{B}^0$ the $\mathcal{B}$-topology on F'. Since every disk $B \in \mathcal{B}$ is compact, hence weakly compact, $\mathcal{B}^0$ is compatible with the duality between F and F' by the Mackey-Arens Theorem (Section 6:2); hence the weak closure of a disk of F' is identical with its closure for the topology $\mathcal{B}^0$. Since F is barrelled, the von Neumann bornology of $E = F'_c$ coincides with the equicontinuous bornology. Thus, if H is a closed and bounded disk in E, then H is a weakly closed equicontinuous set, hence compact for the topology of compact convergence on F' (Proposition (1) of Section 6:1); *a fortiori*, H is compact for $\mathcal{B}^0$ and the space E is, therefore, hypo-Montel.

COROLLARY (1): *If F is a hypo-Montel bornological locally convex space, then its strong dual is a hypo-Montel space.*

Proof: F is sequentially complete (Remark (2)), hence completely bornological (Proposition (1) of Section 4:3) and, therefore, barrelled (Proposition (5) of Section 5:2). Thus F'_c is a hypo-Montel space by Proposition (1). However, F'_c is the strong dual of F since every closed bounded disk in F is compact.

COROLLARY (2): *If F is a Fréchet space then F'_c (cf. Proposition (1)) is a hypo-Montel space. Indeed, F is barrelled.*

REMARK (3): Corollary (2) implies, in particular, that if F is a Banach space, then F'_c is a hypo-Montel space; however, F'_c is not a Montel space in general, since it is barrelled if and only if F has finite dimension (Exercise 7·E.1).

7:2 SCHWARTZ SPACES

7:2·1

DEFINITION (1): (a): *Let E be a separated convex bornological space. A set $A \subset E$ is said to be* BORNOLOGICALLY COMPACT *(b-*COMPACT*) if there exists a bounded disk $B \subset E$ such that A is compact in E_B;*

(b): *A separated convex bornological space is said to be a* SCHWARTZ CONVEX BORNOLOGICAL SPACE, *and its bornology $\mathcal{B}$ is called a* SCHWARTZ BORNOLOGY *(or* BORNOLOGY OF TYPE (S) *for short), if $\mathcal{B}$ has a base of bornologically compact disks;*

(c): *Let, now, E be a separated locally convex space. Then E is called a* SCHWARTZ LOCALLY CONVEX SPACE *if the equicontinuous bornology of E' is of type* (S), *E is called a* CO-SCHWARTZ LOCALLY CONVEX SPACE *if its von Neumann bornology is of type* (S).

7:2·2

REMARK (1): Since in every separated locally convex space E a *b*-compact subset of ${}^bE$ is compact in E, *every co-Schwartz space is hypo-Montel.*

REMARK (2): *Every Schwartz convex bornological space is complete.* In fact, let $\mathcal{B}$ be a base of the bornology of E consisting of b-compact disks; it suffices to show that each $A \in \mathcal{B}$ is completant. But, by definition, there exists a bounded disk $B \subset E$ such that A is compact in the normed space E_B, hence A is completant by virtue of the Corollary to Proposition (1) of Section 3:1.

REMARK (3): It follows from the above Remark (2) that *every co-Schwartz space is bornologically complete.*

7:2˙3 Convergence in Schwartz Spaces

THEOREM (1): *Let E be a regular Schwartz convex bornological space with dual $E^\times$. If (x_n) is a sequence in E which is bounded and convergent to $x \in E$ for $\sigma(E,E^\times)$, then (x_n) converges to x bornologically.*

Proof: Since the sequence (x_n) is bounded in E, the set $A = \{x_n; n \in \mathbb{N}\} \cup \{x\}$ is contained in a b-compact set and hence there is a bounded disk $B \subset E$ such that A is relatively compact in E_B. Denote by $\bar{A}$ the closure of A in E_B. Since the embedding $E_B \to (E,\sigma(E,E^\times))$ is continuous and the topology $\sigma(E,E^\times)$ is separated, E_B and $\sigma(E,E^\times)$ induce the same topology on $\bar{A}$ (*cf.* the proof of Theorem (1) of Section 7:1), hence on A, and the sequence (x_n) must converge to x in E_B.

THEOREM (2): *Let E be a co-Schwartz locally convex space with dual E'. Every sequence (x_n) in E which converges to $x \in E$ for $\sigma(E,E')$ is bornologically convergent to x.*

Proof: The proof is identical to that of Theorem (1), once observed that the sequence (x_n), bounded for $\sigma(E,E')$, is bounded in E (Mackey's Theorem).

7:2˙4 Schwartz Spaces and Reflexivity

Schwartz spaces have very good reflexivity properties.

THEOREM (3): *Every regular Schwartz convex bornological space is reflexive, hence polar.*

Proof: Let E be a regular Schwartz convex bornological space; then the weak topology $\sigma(E,E^\times)$is separated. Since the bornology of E has a base of b-compact disks and every b-compact disk is compact, hence closed for $\sigma(E,E^\times)$, the assertion is an immediate consequence of the Mackey-Arens Theorem (Section 6:2).

COROLLARY: (a): *If E is a regular Schwartz convex bornological space, then $E^\times$ is a Schwartz locally convex space when endowed with its natural topology.*

(b): *If E is a Schwartz locally convex space, then E' is a Schwartz convex bornological space under its equicontinuous bornology.*

Proof: By Theorem (3), $E = (E^\times)'$ bornologically and (a) follow from the definitions, whilst (b) is just a repetition of the defi ition of Schwartz convex bornological spaces.

THEOREM (4): *Every complete Schwartz locally convex space is completely reflexive.*

Proof: Let E be a Schwartz locally convex space and let $(E')^{\times}$ be its bornological bidual. We have to show that $E = (E')^{\times}$ algebraically, hence topologically (Section 6:4). Now E is a topological subspace of $(E')^{\times}$ and is complete, hence closed in $(E')^{\times}$. Thus it is enough to show that E is dense in $(E')^{\times}$. By a Corollary to the Hahn-Banach Theorem (Corollary (3) to Theorem (2) of Section 5:0), this is equivalent to proving that E' is the topological dual of $(E')^{\times}$, since in this case a continuous linear functional on $(E')^{\times}$ vanishing on E must vanish identically on $(E')^{\times}$. Now E' is a regular Schwartz convex bornological space, since $(E')^{\times}$ separates E' by virtue of the fact that E is separated; hence E' is reflexive (Theorem (3)) and, therefore $((E')^{\times})' = E'$.

COROLLARY: *The strong dual of a complete Schwartz locally convex space is completely bornological.*

Proof: In fact, the strong dual of every completely reflexive space is completely bornological (Theorem (1) of Section 6:4).

7:2·5 Intrinsic Characterisation of Schwartz Locally Convex Spaces

Let E be a locally convex space; for every disked neighbourhood U of 0 in E, let (E,U) be the vector space E equipped with the semi-norm p_U, the gauge of U. We denote by E_U the normed space associated with (E,U), i.e. the quotient $E/p_U^{-1}(0)$, where $p_U^{-1}(0) = \{x \in E;\ p_U(x) = 0\}$, endowed with the quotient norm. We also denote by φ_U the canonical continuous linear map of E onto E_U and by $\hat{E}_U$ the Banach space obtained by completing E_U. For every disked neighbourhood V of 0 in E, with $V \subset U$, the identity $(E,V) \to (E,U)$ is continuous and $p_V^{-1}(0) \subset p_U^{-1}(0)$; hence we have a canonical continuous linear map:

$$\varphi_{UV}: E_V \to E_U,$$

obtained from the identity of E by passing to quotients. Clearly the following diagram is commutative:

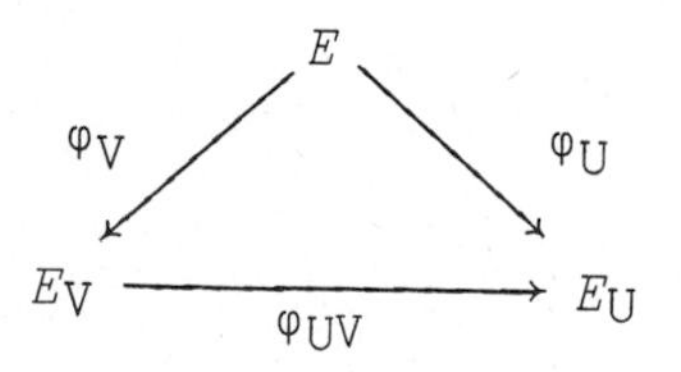

and, consequently, $\varphi_U(V) = \varphi_{UV} \circ \varphi_V(V)$. The set $\varphi_U(V)$ will be called the *CANONICAL IMAGE of V in E_U*.

LEMMA (1): *With the above notation, let E' be the topological dual of E endowed with its equicontinuous bornology. Then U° and V° are equicontinuous disks in E', with $U^\circ \subset V^\circ$, and we have:*

(i): *The dual of the normed space E_U is isometric to $(E')_{U^\circ}$;*

(ii): *If $(E_U)'$ and $(E')_{U^\circ}$ are identified, then the dual map of φ_{UV} is the canonical embedding $(E')_{U^\circ} \to (E')_{V^\circ}$.*

Proof: (i): Let $(E_U)'$ be the topological dual of E_U; with every $u \in (E_U)'$ we associate the element $u \circ \varphi_U \in E'$. The map $u \to u \circ \varphi_U$ is clearly injective and its range is contained in $(E')_{U^\circ} = \bigcup_\lambda \lambda U^\circ$, since u is continuous on E_U and hence bounded on $\varphi_U(U)$ by some $\lambda > 0$. Conversely, if $x' \in (E')_{U^\circ}$, then x' defines a continuous linear functional on (E,U) vanishing on $p_U^{-1}(0)$; in fact, if $x \in p_U^{-1}(0)$, then $x \in \alpha U$ for every $\alpha > 0$ and, since $x' \in \lambda U^\circ$ for a certain $\lambda > 0$, $|\langle x',x\rangle| \leqslant \lambda\alpha$ for all $\alpha > 0$, i.e. $\langle x',x\rangle = 0$. Now x' defines a unique (continuous) linear functional u on E_U via the relation $x' = u \circ \varphi_U$. Therefore, the linear map $u \to u \circ \varphi_U$ is an algebraic isomorphism of $(E')_U$ onto $(E')_{U^\circ}$, hence an isometry in view of the definition of the norms on $(E')_U$ and $(E')_{U^\circ}$.

(ii): The dual of the canonical map φ_{UV} is the map $u \in (E_U)' \to u \circ \varphi_{UV} \in (E_V)'$, which is precisely the canonical embedding of $(E')_{U^\circ}$ into $(E')_{V^\circ}$, once these spaces are identified with $(E_U)'$ and $(E_V)'$ respectively via the maps $u \to u \circ \varphi_U$ and $v \to v \circ \varphi_V$.

THEOREM (5): *A separated locally convex space E is a Schwartz space if and only if it has the following Property: Every disked neighbourhood U of 0 contains a disked neighbourhood V of 0 whose canonical image in E_U is precompact.*

First, we prove the following Lemma.

LEMMA (2): *Let E,F be normed spaces and let $u:E \to F$ be a linear map which maps the unit ball of E onto a precompact subset of F. Then the dual u' of u maps the unit ball of F' onto a compact subset of E'.*

Proof: Obviously u maps every bounded subset of E onto a precompact subset of F, whence denoting by F_p' the dual of F under the topology of precompact convergence, we see that u' is continuous from F_p' to the strong dual E' of E (Theorem (1) of Section 5:5). Now the unit ball of F' is the polar of the unit ball of F and is, therefore, equicontinuous and closed for $\sigma(F',F)$, hence compact in F_p' (Proposition (1) of Section 6:1) and its image unde u' is compact in E'.

Proof of Theorem (5): *Necessity:* Let E be a Schwartz locally convex space and let U be a disked neighbourhood of 0 in E; then U° (which is an equicontinuous disk in E') is contained in an equicontinuous disk of the form V°, with V a disked neighbourhood of 0 in E, such that the embedding $\varphi_{V^\circ U^\circ}:(E')_{U^\circ} \to (E')_{V^\circ}$ maps U°

onto a compact subset of $(E')_{V^\circ}$. By Lemma (2) the image of the unit ball of $((E')_{V^\circ})'$ under the dual map $\varphi'_{V^\circ U^\circ}$ is compact in $((E')_{U^\circ})'$. But $((E')_{V^\circ})'$ and $((E')_{U^\circ})'$ are the (bornological) biduals of E_V and E_U respectively, whence the restriction of $\varphi'_{V^\circ U^\circ}$ to E_V is the canonical map $\varphi_{UV}:E_V \to E_U$ (Lemma (1)). Since E_U is a normed subspace of $((E')_{U^\circ})'$, φ_{UV} maps the unit ball of E_V, hence V, onto a relatively compact subset of the closure $\hat{E}_U$ of E_U in $((E')_{U^\circ})'$. But $\hat{E}_U$ is the completion of the normed space E_U and hence V is precompact in E_U.

Sufficiency: Let A be an equicontinuous disk in E' which we may assume to be of the form $A = U^\circ$, with U a disked neighbourhood of 0 in E. By hypothesis U contains a disked neighbourhood V of 0 in E such that the canonical map $\varphi_{UV}:E_V \to E_U$ maps the unit ball of E_V onto a precompact subset of E_U. The set $B = V^\circ$ is an equicontinuous disk in E' and the dual of φ_{UV}, which by Lemma (1) is the canonical embedding $(E')_A \to (E')_B$, maps A onto a compact subset of $(E')_B$ (Lemma (2)). Thus the equicontinuous bornology of E' is of type (S) and hence E is a Schwartz locally convex space.

COROLLARY (1): *Every bounded subset of a Schwartz locally convex space is precompact.*

Proof: Let E be a Schwartz locally convex space, let A be a bounded subset of E and let W be a disked neighbourhood of 0 in E. We have to prove the existence of a finite set $M \subset E$ such that $A \subset M + W$. Put $U = \frac{1}{2}W$, so that $U + U \subset W$. Since E is a Schwartz space, there is a disked neighbourhood V of 0 in E whose canonical image in E_U is precompact (Theorem (4)). Since A is bounded in E, there exists $\lambda > 0$ such that $A \subset \lambda V$. Let φ be the canonical map of E onto E_U. The set $\varphi(V)$ is precompact in E_U, whence $\lambda\varphi(V) = \varphi(\lambda V)$ is precompact in E_U and we can find a finite subset M of E such that $\varphi(\lambda V) \subset \varphi(M) + \varphi(U)$. It follows that $\varphi(A) \subset \varphi(\lambda V) \subset \varphi(M) + \varphi(U)$; consequently:

$$A \subset \varphi^{-1}(\varphi(A)) = A + p_U^{-1}(0) \subset \varphi^{-1}(\varphi(M)) + \varphi^{-1}(\varphi(U))$$

$$\subset M + U + p_U^{-1}(0) \subset M + U + U \subset M + W,$$

and the precompactness of A follows.

DEFINITION (2): *A FRÉCHET-SCHWARTZ SPACE is a locally convex space which is at the same time a Fréchet space and a Schwartz space.*

COROLLARY (2): *Every Fréchet-Schwartz space E is a Montel space.*

Proof: As a Fréchet space, E is barrelled; it is also hypo-Montel, since its bounded sets are precompact, hence relatively compact (E is complete).

7:3 Silva Spaces

7:3·1

DEFINITION: *A SILVA SPACE is a separated convex bornological space E which is the bornological inductive limit of an increasing sequence (E_n) of Banach spaces such that the unit ball of E_n is compact in E_{n+1} for all n.*

The sequence (E_n) will be called a *DEFINING SEQUENCE* for E. A very simple example of a Silva space is the space $\mathbb{K}^{(\mathbb{N})}$, the countable direct sum of copies of the scalar field, since $\mathbb{K}^{(\mathbb{N})}$ is the inductive limit of the increasing sequence of finite dimensional spaces $E_n = \mathbb{K}^n$. The following Proposition shows that the dual of a Fréchet-Schwartz space is a Silva space when endowed with its equicontinuous bornology, which is of type (S) and has a countable base.

PROPOSITION (1): *Let E be a separated convex bornological space. The following assertions are equivalent:*

(i): *E is a Silva space;*

(ii): *E is a Schwartz space and its bornology has a countable base;*

(iii): *E is the inductive limit of an increasing sequence (E_n) of normed spaces such that the unit ball of E is relatively compact in E_{n+1} for all n.*

Proof: It is clear that (i) implies (ii). To show that (iii) implies (i) let B_n be the closure in E_{n+1} of the unit ball of E_n; since B_n is compact in E_{n+1}, E_{B_n} is a Banach space. Obviously (B_n) is a base for the bornology of E and B_n is compact in $E_{B_{n+1}}$, since $E_{n+1} \subset E_{B_{n+1}}$ and the embedding $E_{n+1} \to E_{B_{n+1}}$ is continuous. Finally, let us show that (ii) implies (iii). Since E is a Schwartz space, its bornology has a base (A_j) of b-compact disks as well as an increasing countable base $(A'_n)_{n\in\mathbb{N}}$. Let A_{j_1} be a member of the base (A_j); the definition of b-compactness and the fact that (A'_n) is a base imply the existence of an integer n_{j_1} such that A_{j_1} is compact in $E_{A'_{n_{j_1}}}$. Since $A'_{n_{j_1}}$ is bounded in E, there exists $A_{n_{j_1}} \in (A_j)$ such that $A'_{n_{j_1}} \subset A_{n_{j_1}}$; again $A_{n_{j_1}}$ is compact in $E_{A'_{n_{j_2}}}$ for some $A'_{n_{j_2}} \in (A'_n)$ and we may suppose that $A'_{n_{j_1}} \subset A'_{n_{j_2}}$. Thus $A'_{n_{j_1}}$ is compact in $E_{A'_{n_{j_2}}}$. By induction we can construct a sequence (B_k) of bounded subsets of E such that $B_k = A'_{n_{j_k}}$ and B_k is compact in $E_{B_{k+1}}$. Since the sequence (A'_n) is increasing, (B_k) is a base of the bornology of E and (iii) follows.

7:3·2 Properties of Silva Spaces

Silva spaces have many important properties which are essentially contained in Theorems (1,2) below.

THEOREM (1): *Let E be a Silva space. A set $A \subset E$ is bornologically closed if and only if it is closed in the topology of ^tE.*

Proof: Let A be a subset of E and let (E_n) be a defining sequence for E. Since the embeddings $E_n \to {}^tE$ are continuous, if A is closed in ^tE then $A \cap E_n$ is closed in E_n for all n and A is b-closed. Conversely, suppose that $A \cap E_n$ is closed in E_n for all $n \in \mathbb{N}$ and let $x \in E$, $x \notin A$. We have to prove the existence of a bornivorous disk $Q \subset E$ such that $(x + Q) \cap A = \emptyset$. Let k be a positive integer such that $x \in E_k$ and let B_n be the unit ball of E_n for $n \in \mathbb{N}$. Since $x \notin A \cap E_k$, there exists a positive number λ_k such that $(x + \lambda_k B_k) \cap A = \emptyset$. In E_{k+1} the set $A \cap E_{k+1}$ is closed and the set $x + \lambda_k B_k$ is compact, and $(x + \lambda_k B_k) \cap (A \cap E_{k+1}) = (x + \lambda_k B_k) \cap A = \emptyset$; hence we can find a positive number λ_{k+1} such that $(x + \lambda_k B_k + \lambda_{k+1} B_{k+1}) \cap A = \emptyset$. Inductively, we can construct a sequence $(\lambda_i)_{i \geq k}$ of positive numbers such that $(x + \sum_{i=k}^{p} \lambda_i B_i) \cap A = \emptyset$ for every integer $p \geq k$. Now the set $Q = \bigcup_{p \geq k} \sum_{i=k}^{p} \lambda_i B_i$ is a bornivorous disk in E and $(x + Q) \cap A = \emptyset$.

COROLLARY (1): *Every Silva space is regular.*

Proof: Since E is separated, the subspace $\{0\}$ is b-closed, hence closed in ^tE. Thus ^tE is separated and E is regular.

By virtue of Theorem (3) of Section 7:2, Corollary (1) implies:

COROLLARY (2): *Every Silva space is reflexive, hence polar.*

COROLLARY (3): (a): *If E is a Silva space, then $E^\times$, endowed with its natural topology, is a Fréchet-Schwartz space:*

(b): *If E is a Fréchet-Schwartz space, then E', endowed with its equicontinuous bornology, is a Silva space.*

Proof: (a): If (B_n) is a countable base for the bornology of E, then (B_n°) (polars in $E^\times$) is a base of neighbourhoods of 0 in $E^\times$. Thus $E^\times$ is metrizable and, being complete (Proposition (1) of Section 5:4), is a Fréchet space. Moreover, E is a Schwartz convex bornological space, so that $E^\times$ is a Schwartz locally convex space (Corollary to Theorem (3) of Section 7:2) and, therefore, a Fréchet-Schwartz space.

(b): If (V_n) is a base of neighbourhoods of 0 in E, then (V_n°) (polars in E') is a base for the bornology of E'. Since E is a Schwartz locally convex space, E' is a Schwartz convex bornological space by definition and its bornology has a countable base. Thus E' is a Silva space by Proposition (1).

COROLLARY (4): *Let E be a Silva space and let M be a bornologically closed subspace of E. Every bounded linear functional on M has a bounded linear extension to all of E.*

Proof: Let $u: M \to \mathbb{K}$ be a bounded linear functional on M; its kernel is b-closed in M, hence in E and, consequently, is closed in ${}^tE$ by Theorem (1). Thus the linear functional u is continuous on M for the topology induced by ${}^tE$ and, by the Hahn-Banach Theorem, u can be extended to a continuous linear functional $\tilde{u}$ on all of ${}^tE$. The continuity of $\tilde{u}$ on ${}^tE$ now implies that $\tilde{u}$ is bounded on ${}^{bt}E$ and hence on E.

THEOREM (2): *Every Silva space is a topological convex bornological space.*

Proof: Let E be a Silva space with defining sequence (E_n) and let B be a bounded subset of ${}^{bt}E$. Suppose that $B \not\subset E$ for all n; then for every integer $k > 0$ there exists $x_k \in B$ such that $x_k \notin kB_k$. Since B is bounded in ${}^{bt}E$, the sequence $y_k = x_k/k$ converges to 0 in ${}^tE$. We shall reach a contradiction by constructing a bornivorous disk in E containing no y_k. Since $y_1 \notin B_1$ and B_1 is closed in E_2, there exists a scalar λ_2, with $0 < \lambda_2 \leqslant 1$, such that:

$$y_1 \notin (B_1 + \lambda_2 B_2);$$

a fortiori, $y_1, y_2 \notin (B_1 + \lambda_2 B_2) \cap B_2$. The set $(B_1 + \lambda_2 B_2) \cap B_2$ is compact, hence closed in E_3, so that there exists a scalar λ_3, with $0 < \lambda_3 \leqslant 1$, for which:

$$y_1, y_2 \notin (B_1 + \lambda_2 B_2) \cap B_2 + \lambda_3 B_3;$$

a fortiori:

$$y_1, y_2, y_3 \notin [(B_1 + \lambda_2 B_2) \cap B_2 + \lambda_3 B_3] \cap B_3.$$

In this way we can construct an increasing sequence (D_n) of disks defined as follows:

$$D_1 = B_1, \qquad D_n = (D_{n-1} + \lambda_n B_n) \cap B_n \qquad (n > 1).$$

Setting $\lambda_1 = 1$ we have for all integers $n \geqslant 1$:

$$D_n \supset \lambda_n B_n \qquad \text{and} \qquad y_1, \ldots, y_n \notin D_n.$$

Now it is clear that $\bigcup_{n=1}^{\infty} D_n$ is a bornivorous disk in E containing no y_k and this contradicts the fact that the sequence (y_k) converges to 0 in ${}^tE$.

REMARK (1): Since every Silva space is polar and has a countable base, Theorem (2) is just a particular case of the following general result proved in the Exercises: '*Every polar convex bornolog ical space with a countable base is topological*' (Exercise 6·E.8).

COROLLARY: *Let E be a Silva space. A subset of E is bounded if and only if it is bounded for $\sigma(E,E^\times)$.*

Proof: By Theorem (2) a subset of E is bounded if and only if it is bounded in ${}^tE$ and by Mackey's Theorem (Section 5:3) a subset of E is bounded in ${}^tE$ if and only if it is bounded for $\sigma({}^tE,({}^tE)') = \sigma(E,E^\times)$.

7:3·3 A Surjectivity Theorem for Duals of Silva Spaces

The following surjectivity theorem will prove very useful in the theory of Partial Differential Equations (*cf.* Chapter VIII).

THEOREM (3): (General Surjectivity Theorem): *Let E,F be Silva spaces and let u be a bounded linear map of E into F. We give $u(E)$ the bornology induced by F and denote by $u':F^\times \to E^\times$ the bornological dual of u. If u is a bornological isomorphism of E onto $u(E)$, then u' is surjective.*

Proof: Put $M = u(E)$ and denote by $\bar{u}$ the map u regarded as a bounded linear map of E onto M, with dual map $\bar{u}':M^\times \to E^\times$. Since $\bar{u}$ is a bornological isomorphism, $\bar{u}'$ is a bornological isomorphism for the natural bornologies on $M^\times$ and $E^\times$ (Proposition (5) of Section 5:5) and hence a surjection. Consider the map:

$$f:y' \in F^\times \to y'|_M \in M^\times,$$

which associates with a bounded linear functional on F its restriction to M; f is surjective. In fact, since u is a bornological isomorphism of E onto $u(E)$ and E is complete (Remark (2) of Section 7:2), $u(E)$ is complete and hence b-closed in F (Proposition (1) of Section 3:2). Thus by Corollary (4) to Theorem (1), every bounded linear functional on M has a bounded extension to all of F, for F is a Silva space, and this implies the surjectivity of f. Now it is clear that the following diagram is commutative:

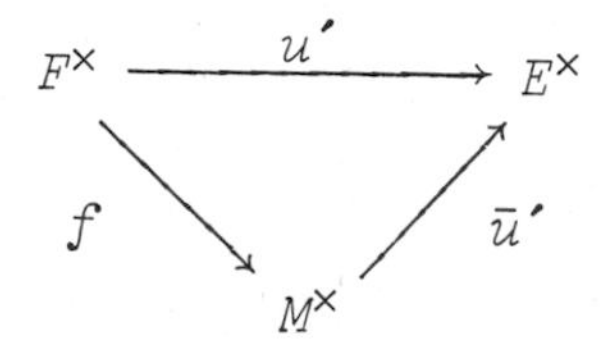

and since $\bar{u}'$ and f are surjections, the Theorem follows.

COROLLARY: *Let E,F be Silva spaces and let $u:E \to F$ be a bounded linear map. Suppose that u is injective and that $u(E)$ is bornologically closed in F. Then the dual $u':F^\times \to E^\times$ of u is surjective.*

Proof: Since $u(E)$ is b-closed in F and F is a Silva space,

$u(E)$ is evidently a Silva space for the bornology induced by F. Thus $u:E \to u(E)$ is a bounded bijection between Silva spaces. But every Silva space is a complete convex bornological space with a countable base, hence u is a bornological isomorphism (Corollary (1) to Theorem (2) of Section 4:4) and the Corollary follows from Theorem (3).

CHAPTER VIII

DISTRIBUTIONS AND DIFFERENTIAL OPERATORS

This final Chapter presents an application of the techniques developed in this book to the solution of partial differential equations. The theorem we shall prove is due to B. Malgrange and is the general existence theorem for solutions, in the space of infinitely differentiable functions, of an arbitrary differential operator with C^∞-coefficients. The choice of this theorem is motivated by the fact that the proof given here mobilises almost all of the fundamental results established in this book. Malgrange's first proof is not a proof of Functional Analysis, since it is based on 'Mittag-Leffler' techniques. Malgrange and Treves have given another proof of Malgrange's Theorem using Functional Analysis, but this proof rests upon the non-elementary theorems of Banach-Dieudonné or Krein-Smulian, and Baire-Banach (*cf.* Bourbaki [*3*]; Chapter IV, §2, n°5, Theorem 5 and Chapter III, §3, n°5, Theorem 3). Our proof is based on the consideration of Silva bornologies, which avoids the use of the above theorems. Moreover, the bornological point of view clarifies the 'true' nature of the notion of a 'convex domain with respect to a differential operator', this property being equivalent to the identity of two natural bornologies (*cf.* Definition (1) and Theorem (1) of Section 8:7).

At the beginning of the Chapter we give a few notions from the theory of distributions which are needed for the statement and proof of Malgrange's Theorem. The reader is referred to H. Hogbe-Nlend [*1*] for a systematic exposition of the theory of distributions from the bornological point of view.

8:0 MULTI-DIMENSIONAL NOTATION

For every integer $n \in \mathbb{N}$ we put:

$$\mathbb{N}^n = \mathbb{N} \times \ldots \times \mathbb{N}, \qquad \mathbb{R}^n = \mathbb{R} \times \ldots \times \mathbb{R} \qquad (n \text{ factors}),$$

and for $\alpha \in \mathbb{N}^n$, $\alpha = (\alpha_1,\ldots,\alpha_n)$ we let $|\alpha| = \sum_{i=1}^{n} \alpha_i$. Denote by $\partial/\partial x_i$ the operator of partial differentiation with respect to the variable x_i, where $x = (x_1,\ldots,x_n) \in \mathbb{R}^n$; then, for every $\alpha \in \mathbb{N}^n$ we put:

$$D^\alpha = \frac{\partial^{|\alpha|}}{\partial x_1^{\alpha_1}\ldots\partial x_n^{\alpha_n}}.$$

Let Ω be a non-empty open subset of $\mathbb{R}^n$. A complex-valued function f on Ω is said to be of class C^∞ or infinitely differentiable on Ω if for every $\alpha \in \mathbb{N}^n$, the partial derivative $D^\alpha f$ exists and is continuous on Ω. The complex vector space of infinitely differentiable functions on Ω is denoted by $C^\infty(\Omega)$. If $f,g \in C^\infty(\Omega)$, then their product $fg \in C^\infty(\Omega)$ and Leibnitz's generalised formula holds:

$$D^\alpha(fg) = \sum_{\beta\leqslant\alpha} \frac{\alpha!}{\beta!(\alpha - \beta)!} D^\beta f D^{\alpha-\beta} g,$$

where:

$$\alpha = (\alpha_1,\ldots,\alpha_n), \quad \beta = (\beta_1,\ldots,\beta_n), \quad \alpha - \beta = (\alpha_1 - \beta_1,\ldots,\alpha_n - \beta_n),$$

$$\alpha! = \prod_{i=1}^{n} \alpha_i!, \quad \beta! = \prod_{i=1}^{n} \beta_i!, \quad (\alpha - \beta)! = \prod_{i=1}^{n} (\alpha_i - \beta_i)!,$$

and $\beta \leqslant \alpha$ means that $\beta_i \leqslant \alpha_i$ for $i = 1,\ldots,n$.

8:1 THE BORNOLOGICAL SPACES $\mathcal{E}(\Omega)$ AND $\mathcal{D}(\Omega)$

8:1·1 The Bornological Space $\mathcal{E}(\Omega)$

A convex bornology may be defined on the vector space $C^\infty(\Omega)$ as follows. A *subset A of $C^\infty(\Omega)$ is said to be* ***BOUNDED*** if for every compact set $K \subset \Omega$ and for every $m \in \mathbb{N}$ we have:

$$\sup_{\varphi\in A} \sup_{\substack{x\in K\\ |\alpha|\leqslant m}} |D^\alpha\varphi(x)| < +\infty.$$

If we let $p_{K,m}(\varphi) = \sup_{\substack{x\in K\\ |\alpha|\leqslant m}} |D^\alpha\varphi(x)|$, then the function $\varphi \to p_{K,m}(\varphi)$ is a semi-norm on $C^\infty(\Omega)$. Thus A is bounded in the above sense if all the semi-norms $p_{K,m}$ are bounded on A when K runs through all compact subsets of Ω and m through all non-negative integers. It is clear that the bounded sets just defined in $C^\infty(\Omega)$ form a *CONVEX BORNOLOGY ON* $C^\infty(\Omega)$ (Example (3) of Section 1:3), and this bornology is separated. Now let (K_j) be an exhaustive sequence of compact subsets of Ω, that is to say, a sequence of compact sets covering Ω and such that $K_j \subset \mathring{K}_{j+1}$. Such a sequence always exists (J. Dieudonné [1], §8). Since every compact subset of Ω is contained in one of the sets K_j, the *sequence* $p_{K_j,m}$ $(j,m \in \mathbb{N})$ *of semi-norms defines the bornology of* $C^\infty(\Omega)$. *Under this bornology (the* C^∞-*BORNOLOGY) the space* $C^\infty(\Omega)$ *will be denoted by* $\mathcal{E}(\Omega)$;

8:1˙2 Metrizability of $\mathcal{E}(\Omega)$

Let us denote by (p_n) the sequence of semi-norms defining the bornology of $\mathcal{E}(\Omega)$; the sequence (p_n) defines also a metrizable, locally convex topology $\mathcal{T}$ on $\mathcal{E}(\Omega)$. Clearly the bornology of $\mathcal{E}(\Omega)$ is the von Neumann bornology of $\mathcal{T}$, hence $\mathcal{T}$ is the topology of ${}^t\mathcal{E}(\Omega)$ and will be called the *CANONICAL TOPOLOGY of* $\mathcal{E}(\Omega)$.

8:1˙3 The Bornological Space $\mathcal{D}(\Omega)$

Let us recall that the support of a complex-valued function f on Ω is the closure (in Ω) of the set $\{x \in \Omega;\ f(x) \neq 0\}$. The support of f is denoted by supp$f$. For every compact set $K \subset \Omega$ let $\mathcal{D}_K(\Omega)$ *be the set of all functions* $f \in C^\infty(\Omega)$ such that supp$f \subset K$. Then $\mathcal{D}_K(\Omega)$ is a vector subspace of $\mathcal{E}(\Omega)$ and hence may be given the bornology induced by $\mathcal{E}(\Omega)$. For $K \subset K'$, $\mathcal{D}_K(\Omega)$ is contained in $\mathcal{D}_{K'}(\Omega)$ and the canonical embedding $\pi_{K'K}:\mathcal{D}_K(\Omega) \to \mathcal{D}_{K'}(\Omega)$ is bounded. Put $\mathcal{D}(\Omega) = \bigcup_K \mathcal{D}_K(\Omega)$, K running through all compact sets in Ω; then $\mathcal{D}(\Omega)$ is a vector space which will be given the inductive limit bornology with respect to the family $\{\mathcal{D}_K(\Omega)\}$. Note that the embedding $\mathcal{D}(\Omega) \to \mathcal{E}(\Omega)$ is bounded for so are all the embeddings $\mathcal{D}_K(\Omega) \to \mathcal{E}(\Omega)$.

8:1˙4 Topological and Bornological Density of $\mathcal{D}(\Omega)$ in $\mathcal{E}(\Omega)$

It can be shown that for every $f \in \mathcal{E}(\Omega)$, there exists a sequence $(\varphi_j) \subset \mathcal{D}(\Omega)$ which converges to f in $\mathcal{E}(\Omega)$ in the topological or bornological sense, these two types of convergence being equivalent by virtue of the metrizability of $\mathcal{E}(\Omega)$ (*cf.* Subsection 8:1˙2).

8:2 DISTRIBUTIONS AS BOUNDED LINEAR FUNCTIONALS

8:2˙1

DEFINITION: *Let* Ω *be a non-empty bounded set in* $\mathbb{R}^n$. *A bounded linear functional on* $\mathcal{D}(\Omega)$ *is called a DISTRIBUTION on* Ω.

Thus the set of distributions on Ω is the *BORNOLOGICAL DUAL of* $\mathcal{D}(\Omega)$. Our definition of distributions is equivalent to the original definition of L. Schwartz [2], and in order to emphasize that we are dealing with the same mathematical concept, we shall make an exception and *denote by* $\mathcal{D}'(\Omega)$ *the space of distributions on* Ω, contrary to our notation for a bornological dual.

It can be shown that every Radon measure on Ω is a distribution on Ω, but that there exist distributions on Ω which are not measures.

For $\varphi \in \mathcal{D}(\Omega)$ and $T \in \mathcal{D}'(\Omega)$ the *value of T at the point* φ will be *denoted by* $\langle T, \varphi \rangle$.

8:2·2 Support of a Distribution

Let Ω_1 be an open subset of Ω. A *DISTRIBUTION T on Ω is said to VANISH on Ω_1* if $\langle T, \varphi \rangle = 0$ for all $\varphi \in \mathcal{D}(\Omega_1)$. If $x_0 \in \Omega$, we say that T *VANISHES IN A NEIGHBOURHOOD of* x_0 if T vanishes on an open neighbourhood of x_0. *The SUPPORT of T, denoted by* supp$T$, *is defined as the complement in Ω of the set of points $x_0 \in \Omega$ such that T vanishes in a neighbourhood of x_0.* It can be shown that this definition generalises the definition of support of a Radon measure and, *a fortiori*, the definition of support of a continuous function.

8:2·3 Distributions with Compact Support

Denoting by $\mathcal{E}'(\Omega)$ the bornological dual of $\mathcal{E}(\Omega)$, one proves the following Theorem:

> *For every bounded linear functional S on $\mathcal{E}(\Omega)$, the restriction of S to $\mathcal{D}(\Omega)$ is a distribution with compact support in Ω, and the map which sends S to its restriction to $\mathcal{D}(\Omega)$ is a linear bijection of $\mathcal{E}'(\Omega)$ onto the set of distributions with compact support in Ω. Thus we may identify $\mathcal{E}'(\Omega)$ with the space of distributions with compact support in Ω.*

For every compact set $K \subset \Omega$ *denote by $\mathcal{E}'(K)$ the space of distributions on Ω with support contained in K*; then clearly, $\mathcal{E}'(\Omega) = \bigcup_K \mathcal{E}'(K)$, K running through all compact subsets of Ω.

8:3 DIFFERENTIAL OPERATORS AND PARTIAL DIFFERENTIAL EQUATIONS

8:3·1 Differentiation and Multiplication Operators

With Ω a non-empty open subset of $\mathbb{R}^n$, let $\alpha \in \mathbb{N}^n$. If $\varphi \in \mathcal{E}(\Omega)$, then $D^\alpha\varphi \in \mathcal{E}(\Omega)$ and the map $D^\alpha : \varphi \to D^\alpha\varphi$ is a bounded linear map of $\mathcal{E}(\Omega)$ into $\mathcal{E}(\Omega)$, called the *DIFFERENTIATION OPERATOR*. It is evident that D^α is also a bounded linear map of $\mathcal{D}(\Omega)$ into $\mathcal{D}(\Omega)$, when $\mathcal{D}(\Omega)$ is given the inductive limit bornology described in Section 8:1.

For every function $f \in \mathcal{E}(\Omega)$ the map $\varphi \to f\varphi$ is also linear and bounded from $\mathcal{E}(\Omega)$ into $\mathcal{E}(\Omega)$ (Leibnitz's formula); it is called the *OPERATOR OF MULTIPLICATION by f* and, clearly, its restriction to $\mathcal{D}(\Omega)$ is a bounded linear map of $\mathcal{D}(\Omega)$ into itself.

8:3·2 Linear Differential Operators

A *LINEAR DIFFERENTIAL OPERATOR* on Ω with C^∞-coefficients is any bounded linear map of $\mathcal{E}(\Omega)$ into itself of the form:

$$P:\varphi \to P\varphi = \sum_{\alpha} a_\alpha D^\alpha \varphi,$$

where $\sum_{\alpha}$ is a finite sum, indexed by $\alpha \in \mathbb{N}^n$, of bounded linear operators of the form $\varphi \to a_\alpha D^\alpha \varphi$, $a_\alpha \in \mathcal{E}(\Omega)$. If the functions a_α are complex constants, P is called a *CONSTANT COEFFICIENT OPERATOR* or a *DIFFERENTIAL POLYNOMIAL*.

EXAMPLES: $P = \sum_{i=1}^{n} \frac{\partial^2}{\partial x_i^2}$ is called the *LAPLACIAN* and is *denoted by* Δ; $P = \frac{\partial}{\partial\tau} - \Delta$ is called the *HEAT OPERATOR* and is a differential operator in $\mathbb{R}^{n+1}$, the general point in this space being denoted by $(x_1,\ldots,x_n,\tau)$.

8:3˙3 The Dual of a Differential Operator

If P is a differential operator on Ω, then the restriction to $\mathcal{D}(\Omega)$ is a bounded linear map of $\mathcal{D}(\Omega)$ into itself by Subsection 8:3˙1, and hence we can consider the dual map of P with respect to the duality between $\mathcal{D}(\Omega)$ and $\mathcal{D}'(\Omega)$. In this way we obtain a linear map:

$$P':\mathcal{D}'(\Omega) \to \mathcal{D}'(\Omega),$$

which is bounded for the natural bornology of $\mathcal{D}'(\Omega)$ (and also continuous for both the natural and the weak topology of $\mathcal{D}'(\Omega)$). *The operator P' is called the* DUAL OF THE DIFFERENTIAL OPERATOR P. It follows from the definitions that P' decreases the support, i.e. if $T \in \mathcal{D}'(\Omega)$, then $\operatorname{supp} P'T \subset \operatorname{supp} T$ and hence the restriction of P' to $\mathcal{E}'(\Omega)$ takes its values in $\mathcal{E}'(\Omega)$. Now P, being a bounded linear map of $\mathcal{E}(\Omega)$ into itself, has also a dual map $\bar{P}^*:\mathcal{E}'(\Omega) \to \mathcal{E}'(\Omega)$ with respect to the duality between $\mathcal{E}(\Omega)$ and $\mathcal{E}'(\Omega)$ and from the density of $\mathcal{D}(\Omega)$ and $\mathcal{E}(\Omega)$ (Section 8.1) it follows that $\bar{P}$ coincides with the restriction of P' to $\mathcal{E}'(\Omega)$.

8:3˙4 The Notion of a Partial Differential Equation

8:3˙4(a)

DEFINITION: *A LINEAR PARTIAL DIFFERENTIAL EQUATION in $\mathcal{E}(\Omega)$ is a linear equation of the form:*

$$Pu = f,$$

where P is a differential operator on Ω, $f \in \mathcal{E}(\Omega)$ is given, and u is an unknown function in $\mathcal{E}(\Omega)$ called the SOLUTION OF THE EQUATION in $\mathcal{E}(\Omega)$.

A LINEAR PARTIAL DIFFERENTIAL EQUATION IN THE SPACE OF DISTRIBUTIONS is a linear equation of the form:

$$P'T = S,$$

with $S \in \mathcal{D}'(\Omega)$ *given and* $T \in \mathcal{D}'(\Omega)$ *unknown.*

8:3·4(b) General Existence Problem

Let $Pu = f$ (resp. $P'T = S$) be a partial differential equation. The general existence problem is the problem of giving necessary and sufficient conditions on P and Ω for the given equation to have a solution for any $f \in \mathcal{E}(\Omega)$ (resp. $S \in \mathcal{E}'(\Omega)$). Clearly this problem is equivalent to that of the surjectivity of the operator $P:\mathcal{E}(\Omega) \to \mathcal{E}(\Omega)$ (resp. $P':\mathcal{D}'(\Omega) \to \mathcal{D}'(\Omega)$). Our goal in the present Chapter is to establish the General Existence Theorem in the space $\mathcal{E}(\Omega)$ by using the techniques developed in this book and in particular, the Surjectivity Theorem for duals of Silva spaces (Theorem (3) of Section 7:3).

8:4 THE SILVA SPACE $\mathcal{E}'(\Omega)$

We have seen in Section 8:1 that $\mathcal{E}(\Omega)$ (Ω a non-empty open set in $\mathbb{R}^n$) is a topological convex bornological space whose associated topology is a metrizable topology defined by the sequence of semi-norms:

$$p_{K,m}(\varphi) = \sup_{\substack{x \in K \\ |\alpha| \leq m}} |D^\alpha \varphi(x)|,$$

when m runs through $\mathbb{N}$ and K through an exhaustive sequence of compact subsets of Ω. Thus the bornological dual $\mathcal{E}'(\Omega)$ of $\mathcal{E}(\Omega)$ is also the topological dual of $\mathcal{E}(\Omega)$ endowed with the above topology and, consequently, the natural bornology on the bornological dual $\mathcal{E}'(\Omega)$ coincides with the equicontinuous bornology on the topological dual $\mathcal{E}'(\Omega)$. In this Section we shall show that $\mathcal{E}'(\Omega)$ is a Silva space whose bornological dual is $\mathcal{E}(\Omega)$, which is equivalent to showing that, from the topological point of view, $\mathcal{E}(\Omega)$ is a Fréchet-Schwartz space.

PROPOSITION (1): *$\mathcal{E}(\Omega)$ is a Fréchet-Schwartz space.*

Proof: (a): First, we prove that $\mathcal{E}(\Omega)$ is a Fréchet space. Let $(f_k)_{k \in \mathbb{N}}$ be a Cauchy sequence in $\mathcal{E}(\Omega)$; for every compact set $K \subset \Omega$, $\alpha = (\alpha_1, \ldots, \alpha_n) \in \mathbb{N}^n$ and $\varepsilon > 0$, there exists an integer $N = N(K, \alpha, \varepsilon)$ such that:

$$\sup_{x \in K} |D^\alpha f_k(x) - D^\alpha f_\ell(x)| \leq \alpha \qquad \text{for } k, \ell > N.$$

It follows that the sequence $(D^\alpha f_k)$ converges to a continuous function g_α uniformly on each compact subset Ω and hence, by a classical result on convergence of differentiable functions, for every $\alpha \in \mathbb{N}^n$ $D^\alpha g_0$ exists and satisfies $D^\alpha g_0 = g_\alpha$. Thus $g_0 \in \mathcal{E}(\Omega)$ and (f_k) converges to g_0 in $\mathcal{E}(\Omega)$.

(b): We now show that $\mathcal{E}(\Omega)$ is a Schwartz locally convex space. For this we shall use the internal characterisation of Schwartz spaces (Theorem (5) of Section 7:2). Let, then, U be a disked neighbourhood of 0 in $\mathcal{E}(\Omega)$ of the form:

$$U = \{f \in \mathcal{E}(\Omega);\ \sup_{\substack{x\in K\\ |\alpha|\leqslant m}} |D^\alpha f(x)| \leqslant 1\},$$

where K is compact in Ω and $m \in \mathbb{N}$. Let $\eta > 0$ be such that the compact set:

$$K' = \{x \in \mathbb{R}^n;\ d(x,K) \leqslant \eta\},$$

is contained in Ω, $d(x,K)$ *denoting the distance of* x *from* K, and put:

$$V = \{f \in \mathcal{E}(\Omega);\ \sup_{\substack{x\in K'\\ |\alpha|\leqslant m+1}} |D^\alpha f(x)| \leqslant 1\}.$$

V is a neighbourhood of 0 in $\mathcal{E}(\Omega)$ and we show that its canonical image V_1 in E_U is precompact. Let us *denote by* $C(K)$ *the Banach space of continuous functions on* K *with supremum norm.* The map $f_1 \in E_U \to (D^\alpha f)_{|\alpha|<m}$ is a normed space isomorphism of E_U *into* the product space $C(K) \times \ldots \times C(K)$ (the number of factors being $\nu = \sum_{|\alpha|\leqslant m} 1$). For every $\alpha \in \mathbb{N}^n$, with $|\alpha| \leqslant m$, $D^\alpha V_1$ is an equicontinuous subset of $C(K)$ since, if $x,y \in K$ are such that $|x - y| \leqslant \eta$, then the segment $[x,y]$ is contained in K' and the Theorem of Finite Increments gives:

$$|D^\alpha f(x) - D^\alpha f(y)| \leqslant \|x - y\|.$$

By Ascoli's Theorem $D^\alpha V_1$ is relatively compact in $C(K)$ and hence V_1 is relatively compact in $C(K) \times \ldots \times C(K)$ (ν factors). Since V_1 is contained in E_U, V_1 is precompact in E_U and the proof is complete.

COROLLARY: *Endowed with its equicontinuous (or natural) bornology,* $\mathcal{E}'(\Omega)$ *is a Silva space whose bornological dual is* $\mathcal{E}(\Omega)$.

Proof: Since $E = \mathcal{E}(\Omega)$ is a Fréchet-Schwartz space, E' is a Silva space under its equicontinuous bornology (Corollary (3) to Theorem (1) of Section 7:3). Since E is complete and Schwartz, it is completely reflexive (Theorem (4) of Section 7:2), i.e. $(E')^\times = E$.

8:5 THE SPACES $\mathcal{E}'(K)$ AND THE BORNOLOGICAL STRUCTURE OF $\mathcal{E}'(\Omega)$

For every compact set $K \subset \Omega$ let $\mathcal{E}'(K)$ be the vector space of distributions on Ω whose support is contained in K. Clearly $\mathcal{E}'(K)$

is a vector subspace of $\mathcal{E}'(\Omega)$.

PROPOSITION (1): *$\mathcal{E}'(K)$ is a bornologically closed subspace of $\mathcal{E}'(\Omega)$.*

Proof: Let (T_n) be a sequence in $\mathcal{E}'(K)$ which converges bornologically to T in $\mathcal{E}'(\Omega)$ for every $\varphi \in \mathcal{E}(\Omega)$, $\langle T_n,\varphi\rangle$ converges to $\langle T,\varphi\rangle$. Choose a $\varphi \in \mathcal{D}(\Omega)$ with support contained in the complement of K in Ω; then $\langle T_n,\varphi\rangle = 0$ because $\operatorname{supp} T_n \subset K$ for all $n \in \mathbb{N}$, hence $\langle T,\varphi\rangle = 0$ and, consequently, $\operatorname{supp} T \subset K$.

COROLLARY: *For every compact set $K \subset \Omega$, $\mathcal{E}'(K)$ is a Silva space when endowed with the bornology induced by $\mathcal{E}'(\Omega)$.*

Proof: $\mathcal{E}'(\Omega)$ is a Silva space (Corollary to Proposition (1) of Section 8:4) and every b-closed subspace of a Silva space is again a Silva space.

From now on we shall assume that $\mathcal{E}'(K)$ always carries the bornology induced by $\mathcal{E}'(\Omega)$, whatever the compact set $K \subset \Omega$.

PROPOSITION (2): *$\mathcal{E}'(\Omega)$ is the bornological inductive limit of its subspaces $\mathcal{E}'(K)$ when K runs through the compact subsets of Ω.*

Proof: It is clear that $\mathcal{E}'(\Omega) = \bigcup_K \mathcal{E}'(K)$ and that, whenever $K_1 \subset K_2$, the canonical embedding $\mathcal{E}'(K_1) \to \mathcal{E}'(K_2)$ is bounded. Therefore, it is enough to prove that every bounded subset of $\mathcal{E}'(\Omega)$ is contained and bounded in one of the spaces $\mathcal{E}'(K)$. Let B be a bounded subset of $\mathcal{E}'(\Omega)$; B is equicontinuous and hence uniformly bounded on a neighbourhood V of 0 in $\mathcal{E}(\Omega)$. By virtue of the semi-norms defining the topology of $\mathcal{E}(\Omega)$ (Section 8:4), we may assume that V has the form:

$$V = \{f \in \mathcal{E}(\Omega);\ p_{K,k}(f) \leqslant 1\},$$

where K is compact in Ω and $k \in \mathbb{N}$. It follows that $B \subset \mathcal{E}'(K)$ and hence the assertion.

8:6 THE GENERAL EXISTENCE THEOREM FOR INFINITELY DIFFERENTIABLE SOLUTIONS

8:6·1 Convexity with Respect to a Bounded Linear Operator on $\mathcal{E}'(\Omega)$

We have seen in the previous Section that $\mathcal{E}'(\Omega)$ is the bornological inductive limit of the spaces $\mathcal{E}'(K)$; hence if u is a bounded linear operator of $\mathcal{E}'(\Omega)$ into itself, then its range is the algebraic inductive limit of the images under u of the spaces $\mathcal{E}'(K)$, i.e.:

$$u(\mathcal{E}'(\Omega)) = \varinjlim_K u(\mathcal{E}'(K)).$$

Thus there are two natural bornologies on $u(\mathcal{E}'(\Omega))$: the bornology induced by $\mathcal{E}'(\Omega)$ and the bornology inductive limit of the bornologies induced by $\mathcal{E}'(\Omega)$ on the subspaces $u(\mathcal{E}'(K))$. In general, these two bornologies are different as will soon be clear. We shall say that the open set Ω is *u-CONVEX* if the two bornologies just considered on $u(\mathcal{E}'(\Omega))$ coincide. Since the bornology inductive limit of the spaces $u(\mathcal{E}'(K))$ is always finer than the bornology induced on $u(\mathcal{E}'(\Omega))$ by $\mathcal{E}'(\Omega)$, to say that Ω is u-convex is equivalent to saying that *every subset of* $u(\mathcal{E}'(\Omega))$ *which is bounded in* $\mathcal{E}'(\Omega)$ *must be contained in one of the spaces* $u(\mathcal{E}'(K))$ and necessarily bounded for the topology induced by $\mathcal{E}'(\Omega)$.

Other usual variations on the notion of convexity of an open set with respect to an operator will be given later (Section 8:9), but now we state a Theorem showing the usefulness of such a notion.

8:6·2 Existence Criterion

THEOREM (1): (General Existence Theorem): *Let P be a differential operator on Ω with infinitely differentiable coefficients and let P' be its dual regarded as a map of $\mathcal{E}'(\Omega)$ into itself. Then the following assertions are equivalent:*

(i): *The map $P:\mathcal{E}(\Omega) \to \mathcal{E}(\Omega)$ is surjective;*

(ii): *The following conditions are satisfied:*

(A): *Ω is P'-convex in the sense of Subsection* 8:6·1;
(B): *For every relatively compact and open subset Ω_1 of Ω and for every function $g \in \mathcal{D}(\Omega_1)$, there exists $f \in \mathcal{E}(\Omega_1)$ such that $Pf = g$ on Ω_1.*

The next two sections are devoted to the proof of Theorem (1).

8:7 PROOF OF THE IMPLICATION (ii) => (i) OF THE GENERAL EXISTENCE THEOREM

LEMMA (1): *Condition* (B) *implies that $P':\mathcal{E}'(\Omega) \to \mathcal{E}'(\Omega)$ is injective.*

Proof: Let $T \in \mathcal{E}'(\Omega)$ be such that $P'T = 0$; if K is the support of T and $\varphi \in \mathcal{E}(\Omega)$ we have to show that $\langle T,\varphi\rangle = 0$. Let Ω_1 be an open relatively compact neighbourhood of K in Ω and let $\psi \in \mathcal{D}(\Omega_1)$ be equal to 1 in a neighbourhood Ω_0 of K. Then $\psi\varphi \in \mathcal{D}(\Omega_1)$ and by Condition (B) there exists $\varphi_1 \in \mathcal{E}(\Omega_1)$ such that $P\varphi_1 = \psi\varphi$ on Ω_1. On Ω_0 we have $\psi\varphi_1 = \varphi_1$ and hence $P(\psi\varphi_1) = P\varphi_1 = \psi\varphi$, which implies that:

$$\langle T,\varphi\rangle = \langle T,\psi\varphi\rangle = \langle T,P(\psi\varphi_1)\rangle = \langle P'T,\psi\varphi_1\rangle = 0.$$

REMARK (1): We have also established that:

$$\langle T,\varphi\rangle = \langle P'T,\psi\varphi_1\rangle \qquad \text{for all } T \in \mathcal{E}'(K).$$

LEMMA (2): *Let K be a compact set in Ω. We give $\mathcal{E}'(K)$ and $P'(\mathcal{E}'(K))$ the bornology induced by $\mathcal{E}'(\Omega)$. Then Condition* (B) *implies that $P':\mathcal{E}'(K) \to P'(\mathcal{E}'(K))$ is a bornological isomorphism.*

Proof: We shall show that $P'(\mathcal{E}'(K))$ is b-closed in $\mathcal{E}'(\Omega)$, from which we deduce that $P'(\mathcal{E}'(K))$ is a Silva space and, consequently, that P', being a bounded linear bijection (Lemma (1)), is a bornological isomorphism (Corollary to Theorem (2) of Section 4:4).

We put $E = \mathcal{E}'(\Omega)$ and show that for every bounded disk $A \subset E$, $P'(\mathcal{E}'(K)) \cap E_A$ is closed in E_A. Let $(P'T_n)$ be a sequence in $P'(\mathcal{E}'(K)) \cap E_A$ which converges to an element S in E_A; we have to prove the existence of a distribution $T \in \mathcal{E}'(K)$ such that $P'T = S$. Now $(T_n) \subset \mathcal{E}'(K)$ and by Remark (1), for every $\varphi \in \mathcal{E}(\Omega)$ there exists $\varphi_1 \in \mathcal{E}(\Omega)$ such that $\langle T_n, \varphi \rangle = \langle P'T_n, \psi\varphi_1 \rangle$. This relation proves that the sequence (T_n) is weakly bounded in E and since E is a Silva space, hence a topological convex bornological space, such a sequence is bounded in E and, therefore, relatively compact in E_B for a suitable bounded disk $B \subset E$. Thus (T_n) has a subsequence which converges to some $T \in E_B$. Since $\operatorname{supp} T_n \subset K$, $\operatorname{supp} T \subset K$ and Remark (1) immediately shows that $P'T = S$.

Proof of the Implication (ii) => (i) *of the General Existence Theorem:* Since $P: \mathcal{E}(\Omega) \to \mathcal{E}(\Omega)$ is the dual of the operator $P': \mathcal{E}'(\Omega) \to \mathcal{E}'(\Omega)$, by virtue of the General Surjectivity Theorem (Theorem (3)) established in Section 7:3, it suffices to show that P' is a bornological isomorphism of $\mathcal{E}'(\Omega)$ onto $P'(\mathcal{E}'(\Omega))$, the latter space carrying the bornology induced by the former. Now by Lemma (2) we have, passing to bornological inductive limits, that $\mathcal{E}'(\Omega) = \varinjlim_K \mathcal{E}'(K')$ is isomorphic via P' to $\varinjlim_K P'(\mathcal{E}'(K))$ and by Condition (A) the latter space is isomorphic to $P'(\mathcal{E}'(\Omega))$.

8:8 PROOF OF THE IMPLICATION (i) => (ii) OF THE GENERAL EXISTENCE THEOREM

Assuming P to be surjective, Condition (B) is evidently satisfied: in fact, if $g \in \mathcal{D}(\Omega_1)$, then $g \in \mathcal{D}(\Omega) \subset \mathcal{E}(\Omega)$ and hence there exists $\varphi \in \mathcal{E}(\Omega)$ such that $P\varphi = g$. Let f be the restriction of φ to Ω_1; then $f \in \mathcal{E}(\Omega_1)$ and $Pf = g$ on Ω_1.

In order to prove that Condition (A) holds let B be a bounded set in $P'(\mathcal{E}'(\Omega))$; we have to show the existence of a compact set $K \subset \Omega$ such that B is contained in $P'(\mathcal{E}'(K))$. Since P is surjective, P' is an injection (hence a bijection) of $\mathcal{E}'(\Omega)$ onto $P'(\mathcal{E}'(\Omega))$. Put $A = (P')^{-1}(B)$ and let $\varphi \in \mathcal{E}(\Omega)$. There exists $\psi \in \mathcal{E}(\Omega)$ such that $\varphi = P\psi$ (P is surjective) and hence, for all $T \in A$:

$$\langle T, \varphi \rangle = \langle T, P\psi \rangle = \langle P'T, \psi \rangle,$$

where $P'T \in B$. Since B is bounded, whence weakly bounded, the above relation shows that $\sup_{T \in A} |\langle T, \varphi \rangle| < +\infty$ and hence that A is weakly bounded. But $\mathcal{E}'(\Omega)$ is a Silva space, hence A is bounded in $\mathcal{E}'(\Omega)$ and, consequently, contained in one of the spaces $\mathcal{E}'(K)$ (Proposition (2) of Section 8:5). Thus $B = P'(P')^{-1}(B) \subset P'(\mathcal{E}'(K))$, which completes the proof.

8:9 EXISTENCE THEOREM FOR PARTIAL DIFFERENTIAL EQUATIONS WITH CONSTANT COEFFICIENTS

8:9˙1

The General Existence Theorem of Section 8:6 gives necessary and sufficient conditions for the existence of solutions in the case of an arbitrary differential operator with C^∞-coefficients. In this Section we turn our attention to constant coefficient operators and show that Condition (B) is automatically satisfied, whilst the notion of P'-convexity introduced in Section 8:6 is equivalent to the classical ones. For this we need the notion of fundamental solution of a differential polynomial.

8:9˙2 Fundamental Solutions

Let $P(D) = \sum_{|\alpha| \leq m} a_\alpha D^\alpha$, $a_\alpha \in \mathbb{C}$, be a differential polynomial on $\mathbb{R}^n$. The dual of $P(D)$ is the operator $P(-D) = \sum_{|\alpha| \leq m} (-1)^{|\alpha|} a_\alpha D^\alpha$, which is again a differential polynomial on $\mathbb{R}^n$. A *FUNDAMENTAL SOLUTION* (or *ELEMENTARY SOLUTION*) *of* $P(D)$ is any distribution E on $\mathbb{R}^n$ satisfying the equation:

$$P(D)E = \delta.$$

Every non-zero differential polynomial has a fundamental solution (*cf.* Appendix).

Let us show that Condition (B) *of Theorem* (1) *of Section* 8:7 *is always verified in the case of differential polynomials:* Let Ω_1 be a relatively compact open subset of Ω, let $g \in \mathcal{D}(\Omega_1)$ and let E be a fundamental solution of $P(D)$. If we denote by f the restriction of the convolution $E*g$ to Ω_1, then $f \in \mathcal{E}(\Omega_1)$ and:

$$P(D)f = P(D)(E*g) = (P(D)E)*g = \delta*g = g \quad \text{on } \Omega_1.$$

It follows that $P(-D):\mathcal{E}'(\Omega) \to \mathcal{E}'(\Omega)$ is always injective (Lemma (1) of Section 8:7).

The General Existence Theorem of Section 8:6 now takes the following form:

8:9˙3

THEOREM (1): *Let* $P(D)$ *be a differential polynomial on* $\mathbb{R}^n$ *and let* Ω *be a non-empty open subset of* $\mathbb{R}^n$. *The following assertions are equivalent:*

(i): *The map* $P(D):\mathcal{E}(\Omega) \to \mathcal{E}(\Omega)$ *is surjective;*

(ii): Ω *is* $P(-D)$*-convex in the sense of Subsection* 8:6˙1;

(iii): *For every compact set* $K_1 \subset \Omega$ *there exists a compact set* $K_2 \subset \Omega$ *such that, whenever* $T \in \mathcal{E}'(\Omega)$ *satisfies* $\operatorname{supp} P(-D)T \subset K_1$, *then* $\operatorname{supp} T \subset K_2$;

(iv): *For every compact set* $K_1 \subset \Omega$ *there exists a compact set* $K_2 \subset \Omega$ *such that, whenever a function* $\varphi \in \mathcal{D}(\Omega)$ *satisfies* $\operatorname{supp} P(-D)\varphi \subset K_1$, *then* $\operatorname{supp}\varphi \subset K_2$.

Proof: By virtue of the General Existence Theorem (Theorem (1) of Section 8:6) and the fact that Condition (B) is always satisfied (Subsection 8:9˙2), assertions (i,ii) are equivalent. We shall prove the following implications: (i) => (iv) => (iii) => (ii).

(i) => (iv): Let K_1 be compact in Ω and put:

$$\Phi(K_1) = \{f \in \mathcal{D}(\Omega);\ \operatorname{supp} P(-D)f \subset K_1\}.$$

$\Phi(K_1)$ is a vector subspace of $\mathcal{E}'(\Omega)$ and we wish to prove the existence of a compact set $K_2 \subset \Omega$ such that $\Phi(K_1) \subset \mathcal{E}'(K_2)$. On $\Phi(K_1)$ we consider the norm:

$$\|f\| = \int_{K_1} |P(-D)f|\,dx,$$

(this is indeed a norm, since $P(-D)$ is injective). The canonical embedding of $\Phi(K_1)$ into $\mathcal{E}'(\Omega)$ is bounded: in fact, if B is the unit ball of $\Phi(K_1)$, then $P(-D)B$ is bounded in $L^1(K_1)$ (the *space of integrable functions on* K_1) and hence bounded in $\mathcal{E}'(\Omega)$. Let $\varphi \in \mathcal{E}(\Omega)$ and let $\psi \in \mathcal{E}(\Omega)$ be such that $P\psi = \varphi$; we have:

$$\langle T,\varphi\rangle = \langle T, P(D)\psi\rangle = \langle P(-D)T,\psi\rangle,$$

for all $T \in B$, hence B is weakly bounded and, consequently, bounded in $\mathcal{E}'(\Omega)$. Thus there exists a compact subset K_2 of Ω such that $B \subset \mathcal{E}'(K_2)$ and, therefore, $\Phi(K_1) \subset \mathcal{E}'(K_2)$.

(iv) => (iii): We give a proof by regularisation. Let K_1 be a compact set in Ω, (ρ_ε) a regularising family in $\mathcal{D}(\mathbb{R}^n)$ and K a compact neighbourhood of K_1 in Ω. If $T \in \mathcal{E}'(\Omega)$ and $\operatorname{supp} P(-D)T \subset K_1$, then there exists an $\eta > 0$ such that $\operatorname{supp}(T*\rho_\varepsilon) \subset \Omega$ and $\operatorname{supp}(\rho_\varepsilon * P(-D)T) \subset K$ for all $\varepsilon < \eta$. It follows that $T*\rho_\varepsilon$ is a function in $\mathcal{D}(\Omega)$ such that $\operatorname{supp}(P(-D)(T*\rho_\varepsilon)) \subset K$. Now (iv) implies the existence of a compact set $K_2 \subset \Omega$ such that $\operatorname{supp}(T*\rho_\varepsilon) \subset K_2$ for all $\varepsilon < \eta$ and, letting $\varepsilon \to 0$ we conclude that $\operatorname{supp} T \subset K_2$.

(iii) => (ii): Let A be a bounded subset of $\mathcal{E}'(\Omega)$ contained in $P(-D)\mathcal{E}'(\Omega)$; then A is contained in $\mathcal{E}'(K_1)$ for some compact set $K_1 \subset \Omega$. By (iii) there is a compact set $K_2 \subset \Omega$ such that, if $S = P(-D)T \in A$, then $T \in \mathcal{E}'(K_2)$. Thus $S \in P(-D)\mathcal{E}'(K_2)$ and we conclude that $A \subset P(-D)\mathcal{E}'(K_2)$.

APPENDIX *EXISTENCE OF A FUNDAMENTAL SOLUTION*

We shall prove the following Theorem, which has been used in Section 8:9.

THEOREM (1): *Every non-zero differential polynomial on* $\mathbb{R}^n$ *has a fundamental solution.*

The proof we give is due to B. Malgrange and relies upon the following three Lemmas:

LEMMA (1): *Let* $f(\lambda)$ *be an entire function of a complex variable* λ *and let* $P(\lambda)$ *be a polynomial of degree* m *in which the coefficient of the term with degree* m *is* 1. *Then for every* $\lambda \in \mathbb{C}$ *we have:*

$$|f(\lambda)| \leqslant \frac{1}{r^m} \max_{|\lambda-\lambda'|\leqslant 2mr} |P(\lambda')f(\lambda')|.$$

Proof: We can write $P(\lambda) = (\lambda - z_1)\ldots(\lambda - z_m)$ and induction on m reduces the proof to that of the inequality:

$$|f(\lambda)| \leqslant \frac{1}{r} \max_{|\lambda-\lambda'|\leqslant 2r} |f_1(\lambda')|,$$

where $f_1(\lambda') = (\lambda' - z_1)f(\lambda')$. Now this inequality is obvious if $|\lambda - z_1| \geqslant r$, whilst for $|\lambda - z_1| \leqslant r$ the maximum principle gives:

$$|f(\lambda)| \leqslant \max_{|\lambda'-z_1|\leqslant 1} |f(\lambda')| \leqslant \frac{1}{r} \max_{|\lambda'-z_1|\leqslant r} |f_1(\lambda')| \leqslant \frac{1}{r} \max_{|\lambda'-\lambda|\leqslant 2r} |f_1(\lambda')|.$$

LEMMA (2): *Let* $f(\lambda)$ *be the Fourier-Laplace transform of a function* $\varphi \in \mathcal{D}(\mathbb{R})$ *and write* $\lambda = \sigma + ir$, $|||f|||_r = \int |f(\sigma + ir)|\,d\sigma$. *If* $P(\lambda)$ *is as in Lemma* (1), *then there exists a constant* C, *which depends only upon* m *and* r, *such that:*

$$|||f|||_0 \leqslant C\{|||Pf|||_0 + |||Pf|||_r + |||Pf|||_{-r}\} \qquad (r > 0).$$

Proof: Let I be the set of real numbers σ for which $|P(\sigma)| \leqslant 1$ and let J be the complement of I in $\mathbb{R}$; we estimate $\int_I$ and $\int_J$ separately.

(a): $$\int_J |f(\sigma)|\,d\sigma \leqslant \int_J |P(\sigma)f(\sigma)|\,d\sigma \leqslant |||Pf|||_0.$$

(b): For every $\sigma \in \mathbb{R}$ we have from Lemma (1):

$$|f(\sigma)| \leqslant \left(\frac{4m}{r}\right)^m \max_{|\lambda'-\sigma|\leqslant \frac{1}{2}r} |P(\lambda')f(\lambda')| \leqslant \left(\frac{4m}{r}\right)^m \max_{|\tau'|\leqslant \frac{1}{2}r} |P(\lambda')f(\lambda')|,$$

where $\lambda' = \sigma' + i\tau'$. To estimate the right hand side we put $g = Pf$ and we use Cauchy's formula:

$$g(\lambda') = \frac{1}{2\pi i}\int \frac{g(\sigma - ir)}{\lambda' - \sigma + ir}\,d\sigma - \frac{1}{2\pi i}\int \frac{g(\sigma + ir)}{\lambda' - \sigma - ir}\,d\sigma.$$

It follows that:

$$|g(\lambda')| \leqslant \frac{1}{\pi r}\left\{\int |g(\sigma - ir)|\,d\sigma + \int |g(\sigma + ir)|\,d\sigma\right.$$

$$= \frac{1}{\pi r}\{|||g|||_r + |||g|||_{-r}\},$$

and hence:

$$|f(\sigma)| \leqslant \frac{1}{\pi r}\left(\frac{4m}{r}\right)^m \{|||g|||_r + |||g|||_{-r}\}.$$

Now the set I has a finite measure not exceeding $2m$, since if $\sigma \in I$, then $|\sigma - z_i| \leqslant 1$ for at least one of the points z_i; consequently:

$$\int_I |f(\sigma)|\,d\sigma \leqslant \frac{1}{2\pi r}\left(\frac{4m}{r}\right)^{m+1} \{|||g|||_r + |||g|||_{-r}\},$$

which concludes the proof of the Lemma.

In order to state Lemma (3) we introduce the following notation: if $x_1,\ldots,x_n$ stand for the coordinates in $\mathbb{R}^n$ and if $\varphi \in \mathcal{D}(\mathbb{R}^n)$, we denote by $\hat{\varphi}(\lambda_1,\ldots,\lambda_n) = \int \exp(-i \sum_{j=1}^{n} \lambda_i x_i)dx_1\ldots dx_n$ the Fourier-Laplace transform of φ, and write $\lambda_j = \sigma_j + i\tau_j$ ($j = 1,\ldots,n$) and:

$$|||\varphi||| = \int |\hat{\varphi}(\sigma_1,\ldots,\sigma_n)|\,d\sigma_1\ldots d\sigma_n.$$

LEMMA (3): *Let $P(D)$ be a differential polynomial on $\mathbb{R}^n$ of order m in $\partial/\partial x_1$ and suppose that the coefficient of $\partial^m/\partial x_1^m$ is equal to 1. Then there exists a constant C, depending only on m and r, such that:*

$$|||\varphi||| \leqslant C \sup_{|\rho| \leqslant r} |||e^{\rho x_1} P(D)\varphi|||.$$

Proof: If $R(\lambda_1,\ldots,\lambda_n)$ is the Fourier-Laplace transform of $P(D)\delta$ we have:

$$R(\lambda_1,\ldots,\lambda_n) = i^m \lambda_1^m + \sum_{p=1}^{m} \lambda_1^{m-p} R_p,$$

the R_p's being polynomials in $\lambda_2,\ldots,\lambda_n$. The Fourier-Laplace transform of $P(D)\varphi$ is $G = R\hat{\varphi}$ and by Lemma (2) we have, for all $\sigma_2,\ldots,\sigma_n$:

$$\int |\hat{\varphi}(\sigma_1,\ldots,\sigma_n)|d\sigma_1 \leqslant C\int \{|G(\sigma_1,\sigma_2,\ldots,\sigma_n)| + |G(\sigma_1 - ir,\sigma_2,\ldots,\sigma_n)| + |G(\sigma_1 + ir,\sigma_2,\ldots,\sigma_n)|\}d\sigma_1,$$

from which the desired inequality follows by integrating with respect to $\sigma_2,\ldots,\sigma_n$.

Proof of Theorem (1): Let $P(-D)$ be the dual of $P(D)$; since $P(D)$ is non-zero we may assume, performing if necessary a change of variables, that $P(-D)$ is as in Lemma (3). We give $\mathcal{D}(\mathbb{R}^n)$ the norm:

$$\varphi \to \sup_{|\rho|\leqslant r} |||e^{\rho x_1}\varphi|||, \tag{1}$$

where $r > 0$ is fixed. Since $P(-D)$ is a one-to-one map of $\mathcal{D}(\mathbb{R}^n)$ into $P(-D)\mathcal{D}(\mathbb{R}^n)$, we can define a linear functional E_0 on $P(-D)\circ\mathcal{D}(\mathbb{R}^n)$ by means of the relation:

$$\langle E_0, P(-D)\varphi\rangle = \varphi(0) \qquad \text{for all } \varphi \in \mathcal{D}(\mathbb{R}^n).$$

E_0 is bounded on $P(-D)\mathcal{D}(\mathbb{R}^n)$ for the norm (1), since Lemma (3) yields:

$$|\varphi(0)| \leqslant |||\varphi||| \leqslant C \sup_{|\rho|\leqslant r} |||e^{\rho x_1}P(-D)\varphi|||;$$

hence by the Hahn-Banach Theorem E_0 can be extended to a linear functional E on $\mathcal{D}(\mathbb{R}^n)$ bounded for the norm (1). *A fortiori*, E is bounded for the bornology of $\mathcal{D}(\mathbb{R}^n)$ (Section 8:1), hence is a distribution and satisfies:

$$\langle P(D)E,\varphi\rangle = \langle E, P(-D)\varphi\rangle = \varphi(0),$$

for all $\varphi \in \mathcal{D}(\mathbb{R}^n)$. Thus $P(D)E = \delta$ and E is a fundamental solution.

EXERCISES ON CHAPTER I

1·E.1

Let E be a topological vector space and let $(V_i)_{i\in I}$ be a base of neighbourhoods of 0 in E. For every family $(\lambda_i)_{i\in I}$ of scalars put:

$$B\{(\lambda_i)\} = \bigcap_{i\in I} \lambda_i V_i.$$

Show that the sets $B\{(\lambda_i)\}$ form a base for the von Neumann bornology of E.

1·E.2

Let E be a separated locally convex space, let $\mathcal{V}$ be a base of disked neighbourhoods of 0 in E and let $\mathcal{B}$ be a base for the bornology of E. Prove that if $\mathcal{B}\cap\mathcal{V} \neq \emptyset$, then E is a normed space (Kolmogorov's Theorem).

1·E.3

Consider a topological vector space E, and show that for a subset A of E the following Properties are equivalent:

(i): A is bounded in E (in the von Neumann sense);

(ii): Every countable subset of A is bounded;

(iii): For every sequence (x_n) of points of A and for every sequence (λ_n) of positive scalars converging to 0, the sequence $(\lambda_n x_n)$ converges to 0 in E.

1·E.4

If $(X,\mathcal{B})$ is a bornological set, we say that $\mathcal{B}$ is a *BORNOLOGY WITH A COUNTABLE CHARACTER*, or a *KOLMOGOROV BORNOLOGY*, if a subset A of X belongs to $\mathcal{B}$ whenever every countable subset of A belongs to $\mathcal{B}$. Give a simple example of a vector bornology with a countable character which is not the von Neumann bornology of a topological vector space. (Hint: consider the compact bornology of an infinite-dimensional Banach space).

1·E.5

Let E be a locally convex space whose topology is defined by a family $(p_i)_{i\in I}$ of semi-norms. Show that the von Neumann bornology of E coincides with the bornology defined by the family $(p_i)_{i\in I}$.

1·E.6

If E is a metrizable topological vector space, prove that for every sequence $(B_n)_{n\in \mathbb{N}}$ of bounded subsets of E (in the von Neumann sense), there exists a sequence (λ_n) of scalars for which the set $B = \bigcup_{n=1}^{\infty} \lambda_n B_n$ is again bounded (Mackey's Countability Condition). (Hint: Let $(V_j)_{j\in\mathbb{N}}$ be a countable base of circled neighbourhoods of 0 in E; for every $n \in \mathbb{N}$ one can find a sequence $(\alpha_{n,j})_{j\in\mathbb{N}}$ of positive real numbers such that $B_n \subset \alpha_{n,j}V_j$ for all $j \in \mathbb{N}$. Put $\alpha_j = \max_{1\leqslant n\leqslant j} \{\alpha_{n,j}\}$ and $A = \bigcap_{j=1}^{\infty} \alpha_j V_j$. Then for every $n \in \mathbb{N}$ there exists $\mu_n > 0$ such that $\alpha_{n,j} \leqslant \mu_n\alpha_j$ for all $j \in \mathbb{N}$ and hence $B_n \subset \mu_n A$).

1·E.7

Let E be a metrizable locally convex space and let (p_n) be a sequence of semi-norms defining the topology of E. For $x,y \in E$ put:

$$d(x,y) = \sum_{n=1}^{\infty} 2^{-n} \frac{p_n(x-y)}{1+p_n(x-y)} .$$

(a): Show that d is a distance on E such that $d(\lambda x,0) \leqslant |\lambda| \times d(x,0)$ whenever $x \in E$ and $|\lambda| \geqslant 1$;

(b): Show that a sequence $(x_j) \subset E$ converges to 0 if and only if $d(x_j,0) \to 0$;

(c): Deduce a new proof of Proposition (3) of Section 1:4 when E is locally convex.

1·E.8 BORNIVOROUS SETS

A subset P of a bornological vector space E is called *BORNIVOROUS* if it absorbs every bounded subset of E.

(a): Prove that if E is a topological vector space with its von Neumann bornology, then every neighbourhood of 0 is bornivorous, but the converse need not be true.

(b): Prove that in a metrizable topological vector space E, a circled set that absorbs every sequence converging to 0 is a neighbourhood of 0, and hence deduce that every bornivorous subset of E is a neighbourhood of 0.

(c): Let E be a bornological vector space and let $(B_i)_{i \in I}$ be a base for the bornology of E. For every family of non-zero scalars put $P\{(\lambda_i)\} = \bigcup_{i \in I} \lambda_i B_i$. Show that the sets $P\{(\lambda_i)\}$ form a fundamental system $\mathcal{P}$ of bornivorous sets in E in the sense that every bornivorous set contains at least one member of $\mathcal{P}$. Hence, deduce that E possesses a fundamental system of circled bornivorous sets.

(d): Verify the following assertions:

(i): Every bornivorous set contains 0;

(ii): Every finite intersection of bornivorous sets is bornivorous;

(iii): If P is bornivorous and $Q \supset P$, then Q is bornivorous. Hence the collection of all bornivorous subsets of a bornological vector space is a filter.

(e): Let E,F be bornological vector spaces and let $u:E \to F$ be a bounded linear map. Show that the inverse image under u of a bornivorous subset of F is bornivorous in E and deduce from this that every bounded linear functional on E is bounded on some bornivorous subset of E. Show also that if F is separated, then the only linear map u of E into F which is bounded on every bornivorous set is the map $u = 0$.

1·E.9 THE TOPOLOGY DEFINED BY THE BORNIVOROUS SETS

Let E be a bornological vector space. A subset Ω of E is called *BORNOLOGICALLY OPEN* if the set $\Omega - a$ is bornivorous for every $a \in \Omega$. The complement of a bornologically open set is called *BORNOLOGICALLY CLOSED*. Show that the family of all bornologically open sets defines a topology τ on E. τ is called the *MACKEY-CLOSURE* (or *b-CLOSURE*) *TOPOLOGY* (*cf.* Remark (1) of Section 2:12).

1·E.10 BORNOLOGICAL CONVERGENCE AND BORNIVOROUS SETS

For every subset A of a bornological vector space E *denote by* $A^{(1)}$ *the set of bornological limits in* E *of sequences of points in* A.

(a): Show that a set $P \subset E$ is bornivorous if and only if $0 \notin A^{(1)}$, where A is the complement of P in E.

(b): Deduce from (a) that a set $\Omega \subset E$ is bornologically open (Exercise 1·E.9) if and only if the following Property is satisfied: for every $a \in \Omega$ and for every sequence $(x_n) \subset E$ which converges bornologically to a, there exists a positive integer n_0 such that $x_n \in \Omega$ for all $n \geqslant n_0$.

(c): Deduce from (b) that a subset A of E is bornologically closed if and only if $A = A^{(1)}$.

1·E.11 Bornological Convergence for Filters

It is said that a *FILTER* Φ *on a bornological vector space* E *CONVERGES BORNOLOGICALLY TO* 0 if there exists a bounded set $B \subset E$ such that:

$$\Phi \supset \{\lambda B;\ \lambda \in \mathbb{K},\ \lambda \neq 0\}.$$

Of course, Φ will converge bornologically to x if the filter $\Phi - x$ converges bornologically to 0.

(a): Show that a sequence $(x_n) \subset E$ converges bornologically to 0 if and only if the 'Fréchet filter' associated with (x_n) converges bornologically to 0. (*A set* $A \subset E$ *belongs to the FRÉCHET FILTER associated with* (x_n) if A contains a set of the form $\{x_n; n \geq n_0\}$ with $n_0 \in \mathbb{N}$).

(b): Prove that every filter which converges bornologically to 0 in E contains a bounded subset of E, and deduce that the filter of all bornivorous sets converges bornologically to 0 if and only if E contains a bounded bornivorous set.

(c): Let A be a subset of E. Prove that if $x \in E$ is the bornological limit of a filter on A, then x is also the bornological limit of a sequence of points of A.

1·E.12 Examples of Bornologies in Function Spaces: Distributions

(a): Let Ω be an open subset of $\mathbb{R}$ and denote by $\mathcal{E}(\Omega)$ the vector space of all infinitely differentiable complex valued functions on Ω. Define a set $B \subset \mathcal{E}(\Omega)$ to be bounded if for every compact subset K of Ω and for every $m \in \mathbb{N}$, the following holds:

$$\sup_{\varphi \in B}\ \sup_{\substack{x \in K \\ p \leq m}} |\varphi^{(p)}(x)| < +\infty.$$

The collection of all such bounded sets forms a separated convex bornology on $\mathcal{E}(\Omega)$ called the *C^∞-BORNOLOGY*

(b): Prove that the bornology defined in (a) on $\mathcal{E}(\Omega)$ can also be defined via a countable family of semi-norms and use this to deduce that such a bornology is the von Neumann bornology of a metrizable locally convex topology on $\mathcal{E}(\Omega)$. Obtain the result that a sequence (φ_n) converges bornologically to 0 in $\mathcal{E}(\Omega)$ if and only if for every compact $K \subset \Omega$ and integer $p \in \mathbb{N}$, the sequence $(\varphi_n^{(p)})_{n \in \mathbb{N}}$ converges to 0 uniformly on K.

(c): With the above notation, let f be a complex valued function on Ω. The *SUPPORT of* f is defined to be the closure in Ω of the set $\{x \in \Omega;\ f(x) \neq 0\}$ and f is said to have *COMPACT SUPPORT* if its support is compact in Ω. *Denote by $\mathcal{D}(\Omega)$ the vector space of infinitely differentiable complex valued functions on Ω with compact support.* A set $B \subset \mathcal{D}(\Omega)$ is said to be *BOUNDED* if the following two conditions are satisfied:

(i): All functions $\varphi \in B$ have their support contained in the same compact subset K of Ω;

(ii): For every $m \in \mathbb{N}$ we have:

$$\sup_{\varphi \in B} \sup_{\substack{x \in K \\ p \leq m}} |\varphi^{(p)}(x)| < +\infty.$$

In this way a separated convex bornology is defined on $\mathcal{D}(\Omega)$, called the *CANONICAL BORNOLOGY of* $\mathcal{D}(\Omega)$;
A *DISTRIBUTION on* Ω is any bounded linear functional on the space $\mathcal{D}(\Omega)$ equipped with its canonical bornology.

(d): Prove that a sequence (φ_n) converges bornologically to 0 in $\mathcal{D}(\Omega)$ if and only if it satisfies the following conditions:

(i): There exists a compact set $K \subset \Omega$ such that the support of φ_n is contained in K for all $n \in \mathbb{N}$;

(ii): For every $p \in \mathbb{N}$ the sequence $(\varphi_n{}^{(p)})_{n \in \mathbb{N}}$ converges to 0 uniformly on K.

An interpretation of the bornologies of $\mathcal{E}(\Omega)$ and $\mathcal{D}(\Omega)$ as an 'initial bornology' and an 'inductive limit bornology', respectively, can be found in the Exercises on Chapter II.

1·E.13 A CONVERGENCE PROPERTY IN BANACH SPACES

Show that in a Banach space E, every sequence that converges to 0 converges bornologically to 0 when E is given its compact bornology and hence obtain a new proof of the fact that the compact bornology of E is the von Neumann bornology of no vector topology on E if E has infinite dimension.

1·E.14 SEQUENCES CONVERGENT TOPOLOGICALLY AND NOT BORNOLOGICALLY

Let I be the interval $[0,1]$ and let $\mathbb{R}^I$ be the product vector space endowed with the product topology. $\mathbb{R}^I$ is a locally convex space.

(a): Prove that the set of all sequences of strictly positiv real numbers tending to $+\infty$ has the same cardinality as

(b): Let f be a bijection of I onto the set of sequences (λ_n) as in (a). Show that the sequence $(x_n) \subset \mathbb{R}^I$ defined by $x_n(i) = 1/\lambda_n$ converges to 0 topologically but not bornologically.

EXERCISES ON CHAPTER II

2·E.1

Let E be a vector space over $\mathbb{K}$ and let $\mathcal{B}$ be a bornology on E. If $E \times E$ is given the product bornology $\mathcal{B} \times \mathcal{B}$ and $\mathbb{K} \times E$ the product bornology when $\mathbb{K}$ carries its canonical bornology, show that $\mathcal{B}$ is a vector bornology if and only if the maps $(x,y) \to x + y$ of $E \times E$ into E and $(\lambda,x) \to \lambda x$ of $\mathbb{K} \times E$ into E are bounded.

2·E.2

Let I be an infinite indexing set. Prove that on $\mathbb{K}^{(I)}$ the direct sum bornology is strictly finer than that induced by the product bornology of $\mathbb{K}^I$ and exhibit a subset of $\mathbb{K}^{(I)}$ which is bounded in $\mathbb{K}^I$ but not in $\mathbb{K}^{(I)}$.

2·E.3

Let E,F and G be topological vector spaces with E and F metrizable. Prove that a bounded bilinear map u of ${}^b(E \times F)$ into ${}^bF$ is continuous. (Hint: Use Proposition (3) of Section 1:4).

2·E.4

Let $(E_i)_{i \in I}$ be a family of topological vector spaces, let E be a vector space and for every $i \in I$, let $u_i : E \to E_i$ be a linear map.

(a): There exists a coarsest vector topology on E for which all the maps u_i are continuous. Such a topology is called the *INITIAL TOPOLOGY on E for the maps* u_i.

(b): If each E_i is given its von Neumann bornology $\mathcal{B}_i$, show that on E the von Neumann bornology associated with the initial topology is the initial bornology on E for the maps u_i.

(c): Deduce from (b) that if E is a locally convex space and $\Gamma = (p_i)_{i\in I}$ a family of semi-norms defining the topology of E, then the von Neumann bornology of E coincides with the bornology defined by the family Γ.

2·E.5

Let E be a topological vector space and let F be a subspace of E. Denote by E/F the quotient space of E by F endowed with the quotient topology and by $\varphi : E \to E/F$ the canonical map.

(a): Verify that the quotient topology on E/F is a vector topology.

(b): Prove that φ is bounded when E and E/F are given their respective von Neumann bornologies.

REMARK: There exist a Fréchet space E, whose bounded sets are relatively compact, and a closed subspace F of E such that E/F has a bounded subset which is not contained in the closure of the image under φ of any bounded subset of E (*cf.* N. Bourbaki [3], Chapter IV, §5, Exercise 21).

2·E.6

Let $(E_i, f_{ji})_{i,j\in I}$ be an inductive system of vector spaces E_i, each E_i being endowed with a locally convex topology $\mathcal{T}_i$. Let E be the (algebraic) inductive limit of this system and for each $i \in I$, let $f_i : E_i \to E$ be the canonical map.

(a): Denote by $\mathcal{V}$ the family of all absorbent disks V in E such that $f_i^{-1}(V)$ is a neighbourhood of 0 in E_i for each $i \in I$. Show that $\mathcal{V}$ is a base of neighbourhoods of 0 for a locally convex topology $\mathcal{T}$ on E which is the finest amongst all locally convex topologies on E for which the maps f_i are continuous. The topology $\mathcal{T}$ is called the *LOCALLY CONVEX INDUCTIVE LIMIT OF THE TOPOLOGIES* $\mathcal{T}_i$ and the space $(E, \mathcal{T})$ is called the *LOCALLY CONVEX INDUCTIVE LIMIT OF THE SPACES* $(E_i, \mathcal{T}_i)$.

(b): Show that the convex bornology on E which is the inductive limit of the von Neumann bornologies of the spaces $(E_i, \mathcal{T}_i)$ for the maps f_i is finer than the von Neumann bornology of E.

REMARK: There exists an increasing sequence (E_n) of Banach spaces with continuous embeddings $f_n : E_n \to E_{n+1}$, such that the locally convex inductive limit of the E_n's contains a bounded set which is not bounded for the inductive limit bornology with respect to the sequence (E_n). (*Cf.* G. Köthe: *Topological Vector Spaces*. (Springer-Verlag, Berlin), (1969)). See, however, Exercise 4·E.10.

2·E.7

Let $(E_i)_{i\in I}$ be a family of locally convex spaces, let E be the (algebraic) direct sum of the E_i's and let $f_i : E_i \to E$ $(i \in I)$ be the canonical embedding.

(a): Denote by $\mathcal{V}$ the family of all absorbent disks V in E such that, for each $i \in I$, $f_i^{-1}(V)$ is a neighbourhood of 0 in E_i. Show that $\mathcal{V}$ is a base of neighbourhoods of 0 for a locally convex topology $\mathcal{T}$ on E which is the finest amongst all locally convex topologies on E for which the maps f_i are continuous. $\mathcal{T}$ is called the *LOCALLY CONVEX DIRECT SUM OF THE TOPOLOGIES of the spaces* E_i and $(E,\mathcal{T})$ is called the *LOCALLY CONVEX DIRECT SUM OF THE SPACES* E_i.

(b): Show that the von Neumann bornology of $(E,\mathcal{T})$ is the bornological direct sum of the von Neumann bornologies of the spaces E_i.

2·E.8 THE M-CLOSURE (OR b-CLOSURE) PROPERTY

A convex bornological space E is said to have the M-CLOSURE PROPERTY if for every subset A of E, $A^{(1)} = \bar{A}$, where $A^{(1)}$ is the set of all bornological limits in E of sequences from A and $\bar{A}$ is the bornological closure of A in E. Prove that a separated convex bornological space with a countable base has the M-closure property if and only if it is a normed space.

2·E.9

Let Ω be an open subset of $\mathbb{R}$ and let $\mathcal{E}(\Omega)$ be the convex bornological space constructed in Exercise 1·E.12. For every compact subset K of Ω and for every $p \in \mathbb{N}$ we *denote by* D_K^p *the map* $f \to f^{(p)}|_K$, i.e. *the restriction to K of the p-th derivative of f*. D_K^p maps $\mathcal{E}(\Omega)$ into the Banach space $C(K)$ of continuous functions on K with the supremum norm. Prove that the C^∞-bornology of $\mathcal{E}(\Omega)$ (Exercise 1·E.12) is the initial bornology for the maps D_K^p.

2·E.10

Let Ω be an open subset of $\mathbb{R}$. For every compact subset K of Ω we *denote by* $\mathcal{D}_K(\Omega)$ *the space of infinitely differentiable complex valued functions on* Ω *with support in K* and we give $\mathcal{D}_K(\Omega)$ the bornology induced by $\mathcal{E}(\Omega)$ (Exercise 2·E.9). Prove that the space $\mathcal{D}(\Omega)$ under its canonical bornology (Exercise 1·E.12) is the bornological inductive limit of the spaces $\mathcal{D}_K(\Omega)$, where K runs through the directed set of all compact subsets of Ω and where the maps involved are the canonical embeddings $\mathcal{D}_K(\Omega) \to \mathcal{D}_{K'}(\Omega)$ for $K \subset K'$.

EXERCISES ON CHAPTER III

3·E.1

Let E be a separated convex bornological sapce and let A be a b-closed bounded disk in E.

(a): If (x_n) is a Cauchy sequence in E_A which converges in E, then (x_n) converges in E_A.

(b): Let us say that A is *MACKEY-COMPLETE* if every sequence in A which is a Mackey-Cauchy sequence in E (Definition (2) of Section 3:5) is bornologically convergent to an element of A. Prove that every Mackey-complete bounded disk in E is completant.

3·E.2

A separated convex bornological space is said to be *Mackey-complete* if every Mackey-Cauchy sequence in E is bornologically convergent.

(a): Every complete convex bornological space is Mackey-complete.

(b): Let $\mathcal{B}$ be the von Neumann bornology of a separated locally convex space E. Prove that if the space $(E,\mathcal{B})$ is Mackey-complete, then it is complete.

The following Exercise characterises all those convex bornological spaces that are complete whenever they are Mackey-complete.

3·E.3

Let E be a Mackey-complete convex bornological space. If A is a subset of E we call ℓ^1-*HULL of* A, *denoted by* $\hat{\Gamma}(A)$, the set of of convergent series of the form $\sum_{n=1}^{\infty} \lambda_n x_n$, where (x_n) is a sequence in A and (λ_n) is a sequence of scalars such that $\sum_{n=1}^{\infty} |\lambda_n| \leq 1$.

(a): Show that if E is a complete convex bornological space, then the ℓ^1-hull of every bounded set is bounded.

Conversely, if E is Mackey-complete and if the ℓ^1-hull of every bounded subset of E is again bounded, then E is complete. In order to establish this assertion, proceed as follows: Let A be a bounded subset of E and put $B = \hat{\Gamma}(A)$.

(b): Show that every bornologically convergent series of the form $\sum_{n=1}^{\infty} \lambda_n x_n$, where $(x_n) \subset A$ and $\sum_{n=1}^{\infty} |\lambda_n| \leq 1$, converges in E_B.

(c): Prove that B is a completant disk and hence deduce the result stated in (a).

(d): A separated convex bornological space is called *SATURATED* if the b-closure of every bounded set is bounded. Deduce from (a) that every Mackey-complete saturated convex bornological space is complete and hence recover the result of Exercise 3·E.2(b).

3·E.4

Let E be a convex bornological space. It is possible to construct a pair $(i,\tilde{E})$, consisting of a complete convex bornological space $\tilde{E}$ and a bounded linear map $i:E \to \tilde{E}$, with the following Universal Property:

(P): *For every bounded linear map u of E into a complete convex bornological space F, there exists a unique bounded linear map $\tilde{u}:\tilde{E} \to F$ such that $u = \tilde{u} \circ i$.*

(a): Prove that if the pair $(i,\tilde{E})$ exists, it is unique up to bornological isomorphism. The space $\tilde{E}$ is called the *BORNOLOGICAL COMPLETION of E*.

(b): Show that E may be assumed to be separated. Let, then, $E = \varinjlim(E_i,\pi_{ji})$ be a representation of E as a bornological inductive limit of normed spaces E_i with injective maps π_{ji}. Let $\hat{E}_i$ be the completion of E_i and for $i \leq j$ let $\hat{\pi}_{ji}: \hat{E}_i \to \hat{E}_j$ be the canonical extension of π_{ji} to the completions.

(c): Show that $(\hat{E}_i,\pi_{ji})$ is an inductive system of convex bornological spaces, whose bornological inductive limit will be denote by $\check{E} = \varinjlim(\hat{E}_i,\pi_{ji})$.

(d): Denote by $\tilde{E}$ the separated convex bornological space associated with $\check{E}$ and show that $\tilde{E}$ is complete. If i is the composition of the canonical maps $E \to \check{E}$ and $\check{E} \to \tilde{E}$, then $(i,\tilde{E})$ is the required pair.

(e): The map i is injective if and only if for every $x \in E$, $x \neq 0$, there exists a bounded linear map u of E into a complete convex bornological space such that $u(x) \neq 0$.

3·E.5

Let $E = \mathcal{P}_0$ be the vector space of all polynomials in the real variable x that vanish at the origin. Define a set $B \subset \mathcal{P}_0$ to be bounded if there exist two positive reals ε and M such that $|p(x)| \leqslant M$ whenever $|x| \leqslant \varepsilon$ and $p \in B$.

(a): Show that the family $\mathcal{B}$ of all bounded subsets of $\mathcal{P}_0$ is a convex bornology having as a base the sequence (B_n) defined by:

$$B_n = \{p \in \mathcal{P}_0;\ |p(x)| \leqslant 1 \text{ for } x \in [-1/n, 1/n]\}.$$

(b): Show that the gauge of B_n is the uniform norm on $[-1/n, 1/n]$. Hence the completion $\hat{E}_{B_n}$ of E_{B_n} is the space of continuous functions on $[-1/n, 1/n]$ vanishing at the origins.

(c): Put $\check{E} = \varinjlim(\hat{E}_{B_n}, \hat{\pi}_{mn})$, where $\hat{\pi}_{mn}: \hat{E}_{B_n} \to \hat{E}_{B_m}$ is the extension of the canonical embedding $E_{B_n} \to E_{B_m}$ (see Exercise 3·E.4(b)). Prove that a bounded linear map of E into a complete convex bornological space is identically zero, and hence deduce that the bornological completion of $\mathcal{P}_0$ reduces to $\{0\}$.

EXERCISES ON CHAPTER IV

4·E.1 A NON-TOPOLOGICAL BORNOLOGY

Let E be a Banach space and let $\mathcal{K}$ be its compact bornology. Show that if E has infinite dimension, then there is no separated locally convex bornology on E whose von Neumann bornology coincides with $\mathcal{K}$ (see Exercises 1·E.4,13) and hence that the compact bornology of an infinite-dimensional Banach space is not a topological bornology.

4·E.*2 A NON-BORNOLOGICAL TOPOLOGY

The following is really an exercise on Chapter V but is given here to illustrate the symmetry between topology and bornology.

Let E be a non-reflexive Banach space and let E' be its dual endowed with the weak topology $\sigma(E',E)$. Show that there are bounded linear functionals on E' that are not continuous and hence that $\sigma(E',E)$ is not a bornological topology.

4·E.3 A BORNOLOGICAL TOPOLOGY WHICH IS NOT COMPLETELY BORNOLOGICAL

Let $E = \mathbb{R}^{(\mathbb{N})}$ be the space of real sequences with only finitely many non-zero terms. Consider, for example, the following norm on E:

$$||x|| = \sum_{n=1}^{\infty} |x_n| \qquad \text{if } x = (x_n) \in E.$$

Then the topology defined by this norm on E is bornological, but not completely bornological. (Hint: Use the Closed Graph Theorem).

4·E.4 PERMANENCE PROPERTIES OF BORNOLOGICAL OR COMPLETELY BORNOLOGICAL TOPOLOGIES

Let $(E_i)_{i \in I}$ be a family of locally convex spaces, let E be a vector space, and for every $i \in I$ let $u_i : E_i \to E$ be a linear map.

(a): Denote by $\mathcal{V}$ the family of all absorbent disks V in E such that for each $i \in I$, $u_i^{-1}(V)$ is a neighbourhood of 0 in E_i. Prove that $\mathcal{V}$ is a base of neighbourhoods of 0 for a locally convex topology on E, called the *FINAL LOCALLY CONVEX TOPOLOGY for the maps* u_i.

(b): If all the spaces E_i are bornological (resp. completely bornological), then E, when endowed with the final locally convex topology, is bornological (resp. completely bornological if E is separated).

(c): Deduce from (b) that a quotient of a bornological (resp. completely bornological) locally convex space is again bornological (resp. completely bornological).

4·E.5 CHARACTERISATIONS OF BORNOLOGICAL TOPOLOGIES

Let $\eta = (\eta_n)$ be a sequence of strictly positive real numbers tending to $+\infty$. A sequence (x_n) in a bornological vector space E is said to be η-*DECREASING* if the sequence $(\eta_n x_n)$ is bounded in E.

(a): Prove that every η-decreasing sequence converges bornologically to 0 and give an example of a sequence that converges bornologically to 0 without being η-decreasing.

(b): Show that the conclusion of Lemma (1) of Section 4:2 still holds if u maps η-decreasing sequences in E onto bounded sequences in F.

(c): Obtain a new characterisation of bornological locally convex spaces improving all the characterisations given in Theorem (1) of Section 4:2.

4·E.6 INTERNAL CHARACTERISATIONS OF BORNOLOGICAL TOPOLOGIES

(a): Establish the following result: In a bornological vector space E, every disk that absorbs all η-decreasing sequences (Exercise 4·E.5) is bornivorous.

(b): Use (a) to obtain the following *internal characterisations* of bornological locally convex spaces:

For a locally convex space E the following assertions are equivalent:

(i): E is bornological;

(ii): Every bornivorous disk in E is a neighbourhood of 0;

(iii): Every disk that absorbs the compact subsets of E is a neighbourhood of 0;

(iv): Every disk in E that absorbs all sequences which converge bornologically to 0 is a neighbourhood of 0;

(v): Every disk in E that absorbs all η-decreasing sequences is a neighbourhood of 0.

4·E.7 INTERNAL CHARACTERISATIONS OF COMPLETELY BORNOLOGICAL TOPOLOGIES

Let E be a separated locally convex space and let E_0 be the complete convex bornological space associated with ${}^bE$. Establish the equivalence of the following assertions:

(i): E is completely bornological;

(ii): Every disk in E that absorbs all completant bounded disks (i.e. all bounded disks in E_0) is a neighbourhood of 0;

(iii): Every disk in E that absorbs all sequences which converge bornologically to 0 in E_0 is a neighbourhood of 0;

(iv): Every disk in E that absorbs all sequences that are η-decreasing in E_0 is a neighbourhood of 0.

4·E.8

Show that only complete vector bornology on $\mathbb{R}^{(\mathbb{N})}$ is the finite-dimensional bornology and deduce that an infinite-dimensional Banach space cannot have a countable dimension. (Hint: Use the Closed Graph Theorem).

4·E.9 A COMPACT BORNOLOGY WHICH IS NOT CONVEX

Let E be the space $\mathbb{R}^{(\mathbb{N})}$ under the norm:

$$||x|| = \sum_{n=1}^{\infty} |x_n| \qquad \text{if } x = (x_n) \in E.$$

On E the compact bornology and the bornology of compact disks are not the same.

4·E.10 LOCALISATION OF COMPLETANT BOUNDED DISKS

If E is a complete convex bornological space with a countable base, then every completant bounded disk of ${}^{bt}E$ is bounded in E. (Hint: Use the Closed Graph Theorem).

EXERCISES ON CHAPTER V

5·E.1 INFRA-BARRELLED SPACES

Let E be a separated locally convex space with dual E'.

(a): The following assertions are equivalent:

(i): Every strongly bounded subset of E' is equicontinuous;

(ii): Every closed bornivorous disk in E is a neighbourhood of 0.

E is called *INFRA-BARRELLED* or *QUASI-BARRELLED* if it satisfies either of the equivalent properties (i) or (ii). Every separated locally convex space which is barrelled or bornological is evidently infra-barrelled.

(b): Let F be a locally convex space and let H be a family of continuous linear maps of E into F which is equibounded on each subset of E that is bounded for the von Neumann bornologies of E. Show that if E is infra-barrelled, then H is equicontinuous.

(c): Prove that a bornologically complete locally convex space is barrelled if it is infra-barrelled.

5·E.2 STRONGLY BOUNDED AND WEAKLY BOUNDED SETS

Let E and F be separated locally convex spaces, with E bornologically complete. Prove that if H is a family of continuous maps of E into F which is simply bounded, then H is equibounded on each subset of E which is bounded for the von Neumann bornology of E.

5·E.3 COMPLETENESS OF STRONG DUALS

Let E be a regular convex bornological space and let $\mathcal{B}$ be the family of subsets of E defined as follows: $B \in \mathcal{B}$ if there exists a sequence (x_n) in E which converges bornologically to 0, such that B is contained in the disked hull of the sequence (x_n).

(a): Show that $\mathcal{B}$ is a vector bornology on E and that E and $(E,\mathcal{B})$ have the same bornological dual, denoted by $E^\times$.

(b): Prove that $E^\times$, endowed with the $\mathcal{B}$-topology, is a complete locally convex space.

(c): Hence obtain the result that the topological dual of a bornological separated locally convex space is complete for the topology of uniform convergence on the sequences that converge bornologically to 0.

5•E.4 EXTERNAL DUALITY BETWEEN BORNOLOGICAL TOPOLOGY AND TOPOLOGICAL BORNOLOGY

Let E be a regular convex bornological space. Show that if $E^\times$ is a bornological locally convex space under its natural topology, then E is topological.

5•E.5 EXTENSION OF BOUNDED LINEAR FUNCTIONALS AND HAHN-BANACH THEOREM

(a): Let E be a separated locally convex space and let F be a subspace of E endowed with the bornology induced by ${}^bE$. Show that every bounded linear functional on F has a bounded extension to all of E if and only if the family of intersections of bornivorous disks in E with F defines a semi-bornological topology on F (for the definition of this topology see Exercise 6•E.2). Hence obtain some examples of locally convex spaces such that every bounded linear functional on a subspace has a bounded extension to the whole space.

(b): Give an example of a complete convex bornological space with a countable base in which a bounded linear functional on a b-closed subspace has no bounded extension to the whole space (*cf.* Exercise 3•E.5).

(c): Let E be a separated convex bornological space in which every b-closed subspace is also closed for ${}^tE$. Show that every bounded linear functional on a b-closed subspace of E can be extended to a bounded linear functional on all of E. A regular convex bornological space with this property is called a *HAHN-BANACH SPACE* or an (HB)-*SPACE* for short.

(d): Show that, conversely, every b-closed subspace of an (HB)-space E is closed in ${}^tE$.

(e): Prove the following assertions:

(i): If E is an (HB)-space, then ${}^{bt}E$ is an (HB)-space;

(ii): Every b-closed subspace of an (HB)-space is again an (HB)-space.

(iii): Every separated quotient of an (HB)-space is an (HB)-space;

(iv): If $I = [0,1]$, then the bornological product $\mathbb{K}^I$ is not an (HB)-space. Deduce that if I is an index set, then the bornological product $E = \prod_{i \in I} E_i$ of a family of normed spaces is an (HB)-space if and only if I is at most countable;

(v): Let $E = \mathbb{K}^{(\mathbb{N})}$ be a bornological direct sum of countably many copies of the scalar field and let F be a Banach space. Then the bornological direct sum $E \oplus F$ of E and F is an (HB)-space;

(iv): There exists, on every separated convex bornological space E, a convex bornology $\mathcal{B}$ such that $(E,\mathcal{B})$ is an (HB)-space.

REMARK: It can be shown that on every separated convex bornological space E, there exists a convex bornology $\mathcal{B}$ such that $(E,\mathcal{B})$ is regular but not an (HB)-space.

EXERCISES ON CHAPTER VI

6·E.1

Show that every infra-barrelled locally convex space (Exercise 5·E.1) is a Mackey space.

6·E.2 MACKEY SPACES AND BORNOLOGICAL LOCALLY CONVEX SPACES

A bornological locally convex space, being infra-barrelled, is a Mackey space (Exercise 6·E.1). In order to characterise the former amongst the latter spaces, show that a separated locally convex space E is bornological if and only if:

(i): E is *SEMI-BORNOLOGICAL*, i.e. every bounded linear functional on E is continuous;

(ii): E is a Mackey space.

6·E.3

Let F be a normed space and let $E = F'_c$ be the dual of E endowed with the $\mathcal{B}$-topology, where $\mathcal{B}$ is the bornology of compact disks of F.

(a): Show that E is reflexive.

(b): Give an example of a normed space F such that $E = F'_c$ is not completely reflexive. (Hint: take a countable direct sum of lines).

6·E.4

Let E be a separated locally convex space such that every strongly bounded sequence in E' is equicontinuous (e.g. E infra-barrelled) and such that E'_β is bornological. Show that if E is reflexive, then it is completely reflexive. In particular, every reflexive normed space is completely reflexive.

6•E.5

Let E be a separated locally convex space and let E' be the dual of E equipped with its equicontinuous bornology.

(a): Prove that a sequence $(x_n') \subset E'$ converges bornologically to 0 if and only if it converges to 0 uniformly on a neighbourhood of 0 in E.

(b): Let $\mathcal{B}$ be the bornology on E' having as a base the disked hulls of sequences that converge bornologically to 0. Show that $\mathcal{B}$ is compatible with the topological dualities $\langle E,E' \rangle$ and $\langle E',(E')^{\times} \rangle$.

(c): Use (b) to show that E is dense in $(E')^{\times}$ when the latter space is given the $\mathcal{B}$-topology, $\mathcal{B}$ being the bornology defined in (b).

6•E.6

Let E be a completely reflexive locally convex space; show that the dual E', when equipped with the equicontinuous bornology, is a reflexive convex bornological space. Conversely, if E is a reflexive convex bornological space, then the bornological dual $E^{\times}$, when given its natural topology, is a completely reflexive locally convex space.

6•E.7

Let E be a complete separated locally convex space. If E' is a reflexive convex bornological space, then E is completely reflexive.

6•E.8

Every polar convex bornological space with a countable base is topological. (Hint: Use Exercise 5•E.4).

6•E.9

Let E be a separated locally convex space. Show that if E is reflexive, then its strong dual is barrelled.

EXERCISES ON CHAPTER VII

7·E.1 HYPO-MONTEL SPACES

Prove the following assertion:

(a): Every hypo-Montel space is reflexive.

(b): Every closed subspace of a hypo-Montel space is hypo-Montel.

(c): If $(E_i)_{i\in I}$ is a family of hypo-Montel spaces, then the product $E = \prod_{i\in I} E_i$ is hypo-Montel.

(d): The strong dual of a Montel space is a Montel space. (Hint: Use Exercise 6·E.9).

(e): If F is a Banach space, then the space $E = F'_c$ (notation as in Proposition (1) of Section 7:1) is barrelled if and only if the dimension of F is finite.

7·E.2 PERMANENCE PROPERTIES OF SCHWARTZ BORNOLOGIES

(a): Let E be a separated convex bornological space and suppose that for every bounded subset A of E there exists a bounded disk $B \subset E$ such that A is relatively compact in E_B. Then E is a Schwartz space.

(b): Prove that the following are Schwartz spaces:

(i): Every b-closed subspace of a Schwartz space;

(ii): Every separated bornological quotient of a Schwartz space;

(iii): Every bornological direct sum of Schwartz spaces;

(iv): Every bornological product of a sequence of Schwartz spaces.

7•E.3 THE COMPACT BORNOLOGY OF A BANACH SPACE

Show that the compact bornology of a Banach space is a Schwartz bornology. (The following result may be assumed: 'For every compact subset A of a Banach space E there exists a sequence (x_n), which converges to 0 in E, such that A is contained in the closed disked hull of (x_n)).

7•E.4 PERMANENCE PROPERTIES OF SCHWARTZ TOPOLOGIES

(In this Exercise 'space' means locally convex space). Prove that the following are Schwartz spaces (for the definitions see Exercises 2•E.4,5,7):

(i): Every subspace of a Schwartz space;

(ii): Every separated quotient of a Schwartz space;

(iii): Every topological product of Schwartz spaces;

(iv): Every locally convex direct sum of a sequence of Schwartz spaces.

7•E.5 SEPARABILITY OF FRÉCHET-SCHWARTZ SPACES

(a): Let E be a Schwartz locally convex space. Show that for every disked neighbourhood U of 0 in E, the space E_U is separable.

(b): Deduce from (a) that every Fréchet-Schwartz space is separable.

7•E.6 PERMANENCE PROPERTIES OF SILVA SPACES

Prove that the following are Silva spaces:

(i): Every b-closed subspace of a Silva space;

(ii): Every separated bornological quotient of a Silva space;

(iii): Every bornological inductive limit of an increasing sequence (E_n) of Silva spaces with injective maps $E_n \to E_{n+1}$;

(iv): Every bornological product of finitely many Silva spaces.

INDEX

We use the following convention: the first and second numerals refer to the chapter and section respectively, whilst the letter E stands for exercises for a particular chapter.

S

T

BIBLIOGRAPHY

The references listed below are those quoted in this book.

N. BOURBAKI

[1] *Théorie des ensembles, Chapitre III.* Hermann, Paris.
[2] *Algèbre linéaire.* Hermann, Paris.
[3] *Espaces vectoriels topologiques.* Hermann, Paris.

G. CHOQUET

[1] *Topologie.* Masson, Paris, (1964).

J. DIEUDONNÉ

[1] *Eléments d'analyse, Vol. I.* Gauthier-Villars, Paris, (1968).
[2] *Eléments d'analyse, Vol. II.* Gauthier-Villars, Paris, (1969).

H. HOGBE-NLEND

[1] *Distributions et bornologie.* Notas do Inst. Mat. Estat. Univ. Sao Paulo, Serie Matematica, No.3, (1973).

L. SCHWARTZ

[1] *Topologie générale et analyse fonctionnelle.* Hermann, Paris.
[2] *Théorie des distributions.* Hermann, Paris.

REFERENCES FOR ADVANCED STUDIES

The standard reference is:

H. HOGBE-NLEND: *Théories des bornologies et applications.* Springer-Verlag, Berlin, (1971).

which contains an essentially complete bibliography up to 1971. After 1971 very many articles on the subject appeared, as well as the following memoirs:

H. BRANDT: *Nukleare b-Raume.* Doctoral thesis. University of Jena, East Germany, (1972).

J.F. COLOMBEAU: *Differentiation et bornologie.* Doctoral thesis. University of Bordeaux I, (1973).

A. FUGAROLAS: *Interpolacion en los espacios bornologicos.* Doctoral thesis. Autonomous University of Madrid, (1973)

G. GALUSINSKI: Espaces de suites à valeurs vectorielles. *Publ. Math. Bordeaux*, **3**, (1973), pp. 000-000; *Publ. Math. Lyon* **10**, (1973), pp. 000-000.

H. GRANGE: *La bornologie de l'ordre.* Thèse (3me cycle). University of Bordeaux I, (1972).

H. HOGBE-NLEND: *Techniques de bornologie en théorie des espaces vectorielles topologiques et des espaces nucléaires.* Lecture Notes in Mathematics Series, Vol. 331. Springer-Verlag, Berlin, (1973). Les fondaments de la théorie spectrale des algèbres bornologiques. *Bol. Soc. Bras. de Matematica*, **3**, No.1, (1972).

C. HOUZEL: *Espaces analytiques relatifs.* University of Nice, (1972).

J.C. LALANNE: *Espaces de suites, nucléarité et bornologie.* Thèse (3me cycle). University of Bordeaux I, (1973).

M. LAZET: *Applications analytiques dans les espaces bornologiques* Lecture Notes in Mathematics Series, Vol. 332. Springer-Verlag, Berlin, (1973).

J.P LIGAUD: *Dimension diamétrale dans les espaces vectoriels topologiques et bornologiques.* Doctoral thesis. University of Bordeaux I, (1973).

V.B. MOSCATELLI: *Contributions to the theory of bornological linear spaces.* Ph.D. thesis. University of London, (1972).

8480-77-25
5-43